D1155063

# METAL–MICROBE INTERACTIONS

**Special Publications of the Society for General Microbiology**

Publications Officer: Dr Duncan E.S.Stewart-Tull, Harvest House, 62 London Road, Reading RG1 5AS, UK

Publisher: Academic Press

1. Coryneform Bacteria
2. Adhesion of Micro-organisms to Surfaces
3. Microbial Polysaccharides and Polysaccharases
4. The Aerobic Endospore-forming Bacteria: Classification and Identification
5. Mixed Culture Fermentations
6. Bioactive Microbial Products: Search and Discovery
7. Sediment Microbiology
8. Sourcebook of Experiments for the Teaching of Microbiology
9. Microbial Diseases of Fish
10. Bioactive Microbial Products 2: Development and Production
11. Aspects of Microbial Metabolism and Ecology
12. Vectors in Virus Biology
13. The Virulence of Escherichia coli
14. Microbial Gas Metabolism
15. Computer-Assisted Bacterial Systematics
16. Bacteria in Their Natural Environments
17. Microbes in Extreme Environments
18. Bioactive Microbial Products 3: Downstream Processing

Publisher: IRL Press

19. Antigenic Variation in Infectious Diseases
20. Nitrification
21. Carbon Substrates in Biotechnology
22. Gene Structure in Eukaryotic Microbes
23. Spatial Organization in Eukaryotic Microbes
24. Bacterial Infections of Respiratory and Gastrointestinal Mucosae
25. Microbial Inoculation of Crop Plants
26. Metal−Microbe Interactions

This book is based on a symposium of the SGM held in April 1988.

SPECIAL PUBLICATIONS OF THE SOCIETY FOR GENERAL MICROBIOLOGY
—————————— VOLUME 26 ——————————

*576.11*
*M564p*

WITHDRAWN

# METAL – MICROBE
# INTERACTIONS

Edited by

## Robert K.Poole

*Biosphere Sciences Division, King's College London,*
*Campden Hill Road, London W8 7AH, UK*

## Geoffrey M.Gadd

*Department of Biological Sciences, The University, Dundee DD1 4HN, UK*

**1989**
Published for the
**Society for General Microbiology**
by

OXFORD UNIVERSITY PRESS
Oxford New York Tokyo

IRL Press
Eynsham
Oxford
England

©Society for General Microbiology 1989

First Published 1989

*All rights reserved by the publisher. No part of this book may be reproduced or transmitted in any form by any means, electronic or mechanical, including photocopying, recording or any information storage and retrieval system, without permission in writing from the publisher.*

*British Library Cataloguing in Publication Data*

Metal—microbe interactions
  1. Metals. Interactions with micro-organisms
  2. Micro—organisms. Interaction with metals
    I. Poole, Robert K.    II. Gadd, Geoffrey M.
    III. Society for General Microbiology
    IV. Series
    546′.3

*Library of Congress Cataloging-in-Publication Data*
Metal—microbe interactions.
   (Special publications of the Society for General Microbiology ; v. 26)
   Based on an April 1988 symposium sponsored by the Cell Biology Group of the Society for General Microbiology.
   Includes index.
   1. Metals—Physiological effect—Congresses.    2. Metals—Metabolism—Congresses.    3. Microorganisms—Physiology—Congresses.    I. Poole, Robert K.    II. Gadd, Geoffrey M.    III. Society for General Microbiology. Cell Biology Group.    IV. Series: Special publications of the Society for General Microbiology : 26.
   QR92.M45M48 1989    576′.11    88—32904

ISBN 0 19 963024 0 (hardbound)
ISBN 0 19 963025 9 (softbound)

Previously announced as:

ISBN 1 85221 202 0 (hardbound)
ISBN 1 85221 198 9 (softbound)

Typeset and printed by Information Printing Ltd, Oxford, England

# Contents

ALLEGHENY COLLEGE LIBRARY

90-273

# Preface

In April 1988, the Cell Biology Group of the Society for General Microbiology held a symposium entitled 'Metal—Microbe Interactions'. Its purpose was to bring together specialists in apparently disparate aspects of microbiology, unified by their expertise in the toxic and beneficial effects of metals on microorganisms and, in turn, the transformations of metals and metal compounds catalysed by microorganisms. The role of this volume in the Special Publications Series is to record and extend the proceedings of the symposium.

Perhaps the most familiar of these interactions is the use of metals as antimicrobial agents to treat fungal and bacterial infections. Several contributors address the chemical and biochemical basis of metal toxicity and describe recent physiological and molecular genetic studies that attempt to understand the resistance defences that microbes mount. Resistance and detoxification mechanisms are diverse and include transformation of the metal to a less toxic form, preventing metal entry or accelerating its efflux, and the binding of accumulated metal by intracellular proteins. These aspects of metal—microbe interactions have considerable biotechnological potential and some already underlie processes for the biosorption of toxic or precious metals by microbial biomass. In a light-hearted vein, 'Daedalus' (*Nature, London, ***332**, pp. 400 and 493) has speculated on further applications. For example, genetic engineering might allow the crossing of conifers with plants resistant to, say, copper that survive in metal-rich soils to yield naturally fungus- and insect-proof timber, from which the metal could ultimately be recovered on ashing. Likewise, the transfer of genes encoding the polyether sodium ionophores from *Streptomyces* to halophyte plants could give plants well adapted to overcome salinity problems in irrigated land or, if transferred to potato, might produce the ideal crop for making salted potato crisps!

Perhaps more remarkable, but well documented here and elsewhere is the ability of chemolithotrophic bacteria to harness the energy from oxidation of Fe(II) and inorganic sulphur compounds, thereby providing a means of liberating metals from sulphidic ores. A further role for microorganisms in natural mineral transformations is in the deposition and precipitation of metallic compounds on cell surfaces and intracellular matrices. These processes have been responsible for mineral formation on a geological scale. Finally, other bacteria use ferric oxide (magnetite) as biological compasses and reduce Fe(III) to Fe(II), thereby contributing to a biological cycle for iron and perhaps providing an alternative redox reaction for supporting respiration in anoxic sediments.

The Group and the Editors thank those who spoke, listened and added to the success of the Symposium with stimulating discussion. The Meetings Staff of the Society provided their customary, excellent support.

We trust that this contribution to the literature will provide a further stimulus to this rapidly developing field and that those new to this area will find the well referenced chapters a useful source of information and a stimulus to contribute to the field of 'Inorganic Microbiology'.

R.K.Poole
G.M.Gadd

# Contributors

T.J.Beveridge
*Department of Microbiology, College of Biological Science, University of Guelph, Guelph, Canada N1G 2W1*

R.P.Blakemore
*Department of Microbiology, University of New Hampshire, Durham, NH 03824, USA*

G.M.Gadd
*Department of Biological Sciences, University of Dundee, Dundee DD1 4HN, UK*

R.B.Frankel
*Department of Physics, California Polytechnic State University, San Luis Obispo, CA 93407, USA*

S.J.Grayston
*Department of Microbiology, University of Sheffield, Sheffield S10 2TN, UK*

M.N.Hughes
*Department of Chemistry, King's College London, Campden Hill Road, London W8 7AH, UK*

R.A.Laddaga
*Bowling Green State University, Bowling Green, OH 43405, USA*

T.K.Misra
*University of Illinois College of Medicine, Chicago, IL 60680, USA*

P.R.Norris
*Department of Biological Sciences, University of Warwick, Coventry CV4 7AL, UK*

R.K.Poole
*Biosphere Sciences Division, King's College London, Campden Hill Road, London W8 7AH, UK*

M.H.Rayner
*Department of Chemistry, Birkbeck College, University of London, Gordon House, 29 Gordon Square, London WC1H 0PP, UK*

P.J.Sadler
*Department of Chemistry, Birkbeck College, University of London, Gordon House, 29 Gordon Square, London WC1H 0PP, UK*

S.Silver
*University of Illinois College of Medicine, Chicago, IL 60680, USA*

M.Wainwright
*Department of Microbiology, University of Sheffield, Sheffield S10 2TN, UK*

C.White
*Department of Biological Sciences, University of Dundee, Dundee DD1 4HN, UK*

# Abbreviations

| | |
|---|---|
| AF | alternating field |
| ARM | anhysteritic remanent magnetism |
| BIM | biologically induced mineralization |
| BOB | boundary-oriented biomineralization |
| DES | diethylstilbestrol |
| DNP | dinitrophenol |
| EDS | energy-dispersive X-ray spectroscopy |
| EDTA | ethylene diamine tetra-acetic acid |
| FPLC | fast protein liquid chromatography |
| HPLC | high-pressure liquid chromatography |
| HRTEM | high-resolution transmission electron microscopy |
| ICP | inductively-coupled plasma |
| LB | Luria-Bertani |
| $M_r$ | relative molecular mass |
| MD | multidomain |
| MT | metallothionein |
| NMR | nuclear magnetic resonance |
| OP | operator/promoter |
| ORF | open reading frame |
| PCF | polycationized ferritin |
| SD | single domain |
| SEM | scanning electron microscopy |
| sIRM | saturation isothermal remanent magnetization |
| SQUID | superconducting quantum interference device |

CHAPTER 1

# Metal mimicry and metal limitation in studies of metal – microbe interactions

MARTIN N.HUGHES[1] and ROBERT K.POOLE[2]

[1]Department of Chemistry and [2]Biosphere Sciences Division, King's College, London, Kensington Campus, Campden Hill Road, London W8 7AH, UK

## Introduction

A fascinating aspect of metal – microbe interactions is the ability of the microbial cell to deal with the range of metals required for growth and function. The supply of metals to the cell depends upon several external factors such as the concentration of a metal in the local environment and also its bioavailability: the latter factor is related in turn to the solubility of the metallic species and to the presence and properties of any ligand to which the metal cation may be complexed. Uptake of a specific metal from the environment is then under the control of the cell, which is able to distinguish particular metals one from another, and to provide specific transport systems for their translocation to certain intracellular sites. Other metal cations are necessarily excluded from these sites. This precise control of metal binding in biology contrasts with a simple chemical situation, where the competition between metal ions for a ligand is normally won by the metal ion which is the strongest Lewis acid of those present. Thus, if a range of biologically important metals were presented with a variety of ligands, then Cu(II), as the strongest Lewis acid, would dominate the competition for ligands. No other metal ion would bind until the concentration of free Cu(II) had been lowered by complexation to a range where other metal ions would be able to compete. Yet, in contrast, the cell is able to select a metal such as Ni(II) or Zn(II) in the presence of Cu(II) and insert it into an appropriate site. This ability to bind one metal from a range of competing cations is a remarkable example of selectivity (Williams, 1981).

In view of this emphasis on selectivity in the interaction of metals and microbes, it may be necessary to justify the setting up of competitions between a native metal ion and a foreign competitor. Is it possible to arrange conditions where a foreign metal can be taken up by the microbial cell, compete effectively with a native metal for its binding site and eventually replace it? What information can be gained from the study of such interactions?

### Objectives underlying studies on metal mimicry in microbes

Investigations of such metal – metal interactions in microbes allows the exploration of the fundamental biochemistry and physiology of metal-dependent processes and the

1

elucidation of specific mechanisms for metal ion toxicity to microorganisms, as discussed in the following paragraphs.

(i) Replacement of the native metal by another may result in perturbation of normal biological function. The extent and nature of this perturbation may be correlated with the different properties of the essential and replacement cations and may lead to useful mechanistic inferences regarding the function of the native metal ion. Such studies could also give valuable insight into the way in which the cell deals with competitions between naturally occurring metal ions under normal physiological conditions.

(ii) The exploration of metal−metal interaction is likely to promote the understanding of the toxic effects of metals towards microorganisms. Such toxicity often involves displacement of native metals from their biological sites.

(iii) Microorganisms provide real biological models for the study of metal−metal inter-actions that are relevant to other areas of research, for example to topics of medical interest such as the mechanisms of metal toxicity and the use of imaging agents. Thus, there has been much interest recently in the competitive interactions between $Fe^{3+}$ and other metal ions in the $+III$ oxidation state. Studies on the competition between $Fe^{3+}$ and $Al^{3+}$ may relate to the transport *in vivo* of aluminium cations and to the neuro-toxicity of aluminium (Edwardson *et al.*, 1986). Similarly studies on competition between $Fe^{3+}$ and $Ga^{3+}$ may be relevant to the antitumour properties of $Ga^{3+}$, which arise from the replacement of iron by gallium in ribonucleotide reductase (Hayes and Hubner, 1983), and to the use of gallium-67 as an imaging agent for a variety of soft tissue tumours and inflammatory abscesses (Hayes, 1978; Welch and Moerlein, 1980), a process which involves the transport of $Ga^{3+}$ bound to the serum iron transport protein transferrin (Vallabhajosula *et al.*, 1980). It is possible that $Ga^{3+}$ may serve generally as a probe of iron transport and function involving $Fe^{3+}$.

(iv) If the native metal ion can be replaced with one that binds similarly at the metal site, then, provided the replacement or 'probe' metal has useful electronic or nuclear properties, the presence of the replacement metal may give useful information about metal-binding sites in terms of site symmetry and the nature of the ligand groups (Hughes, 1981, 1987). This isomorphous replacement approach has been heavily exploited in work on isolated metalloenzymes (Cohn, 1970; Mildvan, 1979) and may well also have applications in work on whole cells.

## Some chemical factors relevant to metal−metal competition

This section gives a brief account of the chemical factors that determine the ability of a 'foreign' metal ion (which may be an essential metal under other circumstances) to compete with a native cation for a biological site. The essential metals and some common toxic metals are shown in *Table 1*, the Periodic Table of the elements.

*The essential metals*

As may be seen in *Table 1*, the essential metals fall into two groups, having distinct properties. The s-block elements $Na^+$ and $K^+$, and $Mg^{2+}$ and $Ca^{2+}$ (Groups IA and

**Table 1.** The Periodic Table of the elements showing essential and toxic metals. Hydrogen, helium, the lanthanides and the actinides are omitted. Essential metals are shown in bold type; toxic metals are italicized.

| Li | Be | | | | | | | | | | | B | C | N | O | F | Ne |
|---|---|---|---|---|---|---|---|---|---|---|---|---|---|---|---|---|---|
| **Na** | **Mg** | | | | | | | | | | | *Al* | Si | P | S | Cl | Ar |
| **K** | **Ca** | Sc | Ti | **V** | **Cr** | **Mn** | **Fe** | **Co** | **Ni** | **Cu** | **Zn** | Ga | Ge | *As* | Se | Br | Kr |
| Rb | Sr | Y | Zr | Nb | **Mo** | Tc | Ru | Rh | Pd | *Ag* | *Cd* | In | *Sn* | *Sb* | Te | I | Xe |
| Cs | Ba | La | Hf | Ta | W | Re | Os | Ir | Pt | Au | *Hg* | *Tl* | *Pb* | Bi | Po | At | Rn |

**Table 2.** Partitioning of metals in ligand matrices

| Oxygen | O/N/S | Nitrogen/sulphur |
|---|---|---|
| Na  Al | V    Pb | Co  Hg |
| K | Cr   Ga | Ni |
| Mg | Mn   Tl | Cu |
| Ca | Fe   Cd | Zn |
|  | Mo |  |

IIA, respectively) are present in relatively high concentrations in biological systems and may be classified as bulk metals. They are distributed selectively, with $K^+$ and $Mg^{2+}$ concentrated inside the cell and $Na^+$ and $Ca^{2+}$ outside the cell. This selective distribution is fundamental to the biological functions of these four cations (Hughes, 1981). The d-block elements are present often at extremely low levels and are usually described as the trace and ultra-trace metals. This group includes the 3d transition metals from vanadium through to zinc and the second row transition metal molybdenum. The d-block elements form complexes much more strongly than do the IA and IIA cations.

### The toxic metals

The well-known toxic metals include cadmium, mercury, lead, tin, thallium and arsenic (a metalloid); silver is less well known but is probably the most toxic element to micro-organisms (Foye, 1977; Friberg *et al.*, 1979). The essential metals may also exert toxic effects if their concentrations are raised to high enough levels. The toxic metals generally bind more strongly to ligands than do the essential metals, so displacing them from their normal sites, though they also exert toxic effects by binding to other sites. They usually bind most strongly to sulphur-containing groups. Recently, however, there has been much concern over a 'new' toxic element, aluminium, which has different chemical characteristics from the familiar toxic metals, and prefers oxygen sites exclusively (Martin, 1986; MacDonald and Martin, 1988).

### Binding groups and selectivity

It is important to consider how metal ions can be bound selectively, as this will give insight into the factors that relate to competitions between metal ions for binding sites.

In general, these three groups of metals, the bulk essential metals, the trace essential metals and the toxic metals, prefer different ligand groupings (Hughes, 1981) (*Table 2*). These preferences for certain ligand environments provide an immediate distinction

3

between the behaviour of individual essential metals that must be incorporated into the design of selective sites. This may be illustrated by further reference to Cu(II), the strongest Lewis acid amongst the essential metals, and to its domination of competition among the trace metals for ligand sites. How could a weak Lewis acid such as Mg(II) even compete against Cu(II)? This is achieved because biological systems present Mg(II) with a ligand domain, involving oxygen donor atoms exclusively, for which Cu(II) has lowest affinity. The binding of Mg(II) is favoured further because it is present in very much higher concentrations than is Cu(II). If the two metals were competing for a ligand environment that included a nitrogen or a sulphur donor atom then the result would be reversed, irrespective of this concentration factor.

Distinction between Cu(II) and other transition metal ions is a more complex problem. The concentration of available Cu(II) must be lowered by providing a ligand that will bind strongly to Cu(II) and which is synthesized in response to the concentration levels of copper. If the bulk of the Cu(II) is strongly complexed by this ligand, then other metals will be able to compete with the residual free Cu(II) because their relative concentrations are now much greater. The metallothioneins are one group of ligands that could bind and effectively remove Cu(II) in this manner. It must then be assumed that Cu(II) can be abstracted from this complex by proteins with sites having high affinity for Cu(II), leading to the eventual incorporation of Cu(II) into copper proteins.

Distinction between the remaining transition metal ions may then be made on the basis of other factors, for example by providing metal-binding sites on proteins which are compatible with a specific geometry that favours one metal ion more than another.

### Choice of replacement metal

The affinity of a metal ion for a particular ligand environment depends upon several factors. For biological activity to be maintained or partially maintained, the replacement metal ion should bind to the site in as similar a fashion as possible as the native metal. To achieve such 'isomorphous' substitution, both native and replacement metal ions should have the same ionic charge and similar ionic radii, and should prefer the same coordination numbers, geometries and ligand types in their coordination complexes. These ligand preferences will vary with the native metal and so may involve sites with 'hard' oxygen or nitrogen donors or those with 'soft' sulphur donors. If metals with higher oxidation states such as Fe(III) are involved, then some negatively charged ligands will probably be included in the coordination shell in order to reduce the overall effective charge on the metal centre. A good guide to the extent of similarity between two cations is provided by a comparison of values of formation constants for their interaction with a range of ligands in solution (Sillen and Martell, 1971).

*Table 3* lists some possible metal-for-metal substitutions, together with ionic radii and comments on the extent of the chemical similarities between the metal cations. The importance of similarity in size must be emphasized; this is usually of greater significance than similarity in charge. The ionic radius of a metal ion is a function of coordination number and this should be borne in mind when making comparisons. *Table 3* includes examples of metal replacement where isomorphous substitution may take place and others where the replacement is unlikely to confer biological activity. In the latter case, the replacement metal ion may still bind at the active site and so allow some conclusions to be drawn regarding the nature of the site by the application of various instrumental

4

**Table 3.** Some metal-for-metal substitutions.

| Native cation | Ionic radius[a] | Replacement | Ionic radius[a] | Chemical comparison | Properties of probe |
|---|---|---|---|---|---|
| $K^+$ | 1.33 | $Tl^+$ | 1.40 | Fairly similar, $Tl^+$ is a soft cation | NMR |
| $Mg^{2+}$ | 0.72 | $Ni^{2+}$ | 0.69 | Similar, except for lack of lability of $Ni^{2+}$ | Paramagnetic |
|  |  | $Mn^{2+}$ | 0.80 | Generally good | Paramagnetic |
| $Ca^{2+}$ | 0.99 | $Mn^{2+}$ | 0.80 | Poor in size and stereo-chemistry | Paramagnetic |
|  |  | $Ln^{3+}$ | $0.93-1.15$ | High charge, binds strongly; moderate overall | Paramagnetic |
|  |  | $Cd^{2+}$ | 0.97 | Competes strongly for sites, soft cation | NMR |
| $Fe^{3+}$ | 0.64 | $Ga^{3+}$ | 0.62 | Good, except for redox reactions | NMR |
|  |  | $Tb^{3+}$ | 1.00 | Moderate, poor in size |  |
|  |  | $Al^{3+}$ | 0.54 | Needs O-donors, generally good, small size causes steric hindrance | NMR |
| $Cu^{2+}$ | 0.69 | $Ca^{2+}$ | 0.99 | Poor (redox reactions!) |  |
| $Cu^+$ | 0.96 | $Ag^+$ | 1.26 | Problems with stereo-chemistry and solubility |  |
| $Zn^{2+}$ | 0.74 | $Co^{2+}$ | 0.72 | Excellent, very successful probe | Paramagnetic |
|  |  | $Cd^{2+}$ | 0.97 | Toxic substitution | NMR |
| Mo |  | W |  | Moderate |  |

[a]Ionic radii in Å (for six-coordination).

techniques and by correlation of toxicity with the properties of the two metal ions. Attention should be drawn to the use of Co(II) to probe Zn(II) sites, which has had many successful applications, often with full maintenance of biological activity, and also to the importance of competition between Cd(II) and both Zn(II) and Ca(II). This is consistent with the ability of Cd(II) to bind with both N/S and O donor groups, as noted in *Table 2*. Cadmium-113 NMR spectroscopy of proteins has proved to be a valuable technique (Summers, 1988).

In all cases, the ability of a replacement metal ion to compete with the native metal can be enhanced by adjustment of their relative concentrations.

## Some complications in studies on metal replacement

Various problems may arise during attempts to persuade microorganisms to accept a replacement metal. Such metal ions may give insoluble products with anions present in the growth medium or in the cell itself. Replacement metals, if taken up by the cell, may still fail to function as effective substitutes for native cations and give partially or completely inactive systems. This may result from incompatible redox properties, or, more generally, because the replacement cation binds to additional sites causing inhibition. The latter situation commonly holds for the toxic, soft metal cations.

An often-unappreciated complication is that the replacement metal ion may be inhibitory because it is unsufficiently labile, that is the rate of exchange of ligands in and out of the metal ion coordination sphere is too slow. The rate at which ligands enter and leave the coordination shell of the metal ion is determined by the polarizing power (charge/radius ratio) of the cation, so it is possible to make simple estimates of the relative lability of different metal ions. The alkali metal ions will be very labile, as will the alkaline earth cations. Other cations, notably those with higher charges, will be less labile. The rates of ligand exchange at some metal cations lie in the sequence $Al^{3+} < Fe^{3+} < Ga^{3+} <<< Mg^{2+} <<< Zn^{2+} < Ca^{2+}$. The lack of lability of certain metal cations explains several phenomena. The cation $Ni^{2+}$ has similar chemistry and ionic radius to $Mg^{2+}$ and could well be regarded as an excellent replacement for $Mg^{2+}$, allowing biological activity to be maintained. However, substitution of magnesium by nickel in enzymes gives inactive products because the Ni(II) centre is not labile enough, and so (for example) may not release the products quickly enough. The enzyme phosphoglucomutase is a rare example where replacement of $Mg^{2+}$ by $Ni^{2+}$ gives an active enzyme (Mildvan, 1977). A more effective replacement for magnesium is manganese, which has been used with great success in the probing of the function of magnesium in enzymes (Cohn, 1970; Mildvan, 1979).

The slowness of ligand exchange rates is especially significant in the case of $Al^{3+}$, and explains why this cation cannot substitute effectively for other cations. Indeed the toxicity of $Al^{3+}$ often results from inhibition of $Mg^{2+}$-dependent enzymes, probably because ligand exchange is about $10^5$ times slower for $Al^{3+}$ than $Mg^{2+}$ (Martin, 1986). Furthermore, $Al^{3+}$ can readily replace $Mg^{2+}$ in biological systems because of its much higher affinity for most ligands. For example, $Al^{3+}$ binds almost $10^7$ times more strongly to $ATP^{4-}$ than does $Mg^{2+}$; this means that in enzymes which form ternary metal ion$-$ATP$-$enzyme complexes, less than nanomolar amounts of free $Al^{3+}$ can compete with millimolar levels of $Mg^{2+}$ (Martin, 1986). It is well known that the $Mg^{2+}$-dependent enzymes yeast and brain hexokinase need activators such as phosphate, citrate or even fluoride. Only recently has it been appreciated that this requirement results from the fact that commercial preparations of ATP are contaminated by aluminium ions, so that the ATP$-Al^{3+}$ complex binds instead of the $Mg^{2+}$ complex to give an inactive enzyme. The function of the anion is merely to remove the $Al^{3+}$ by complexation or precipitation (Womack and Colowick, 1979; Viola *et al.*, 1980).

It should also be noted that replacement metal ions may bind in a slightly different way from the native ion, and so activate by a rather different mechanism. Reference has been made to the important use of $Mn^{2+}$ as a probe for $Mg^{2+}$, despite the significant differences in their ionic radii (see *Table 3*). This could well account for the slightly different biochemical behaviour of these two cations. Thus, the $Mg^{2+}$-activated pyruvate kinase is inhibited by $Li^+$ but the $Mn^{2+}$-activated enzyme is not. Again, there are differences in the catalytic and regulatory properties of the $NAD^+$-specific malic enzyme of *Escherichia coli* depending upon whether the divalent activator is $Mn^{2+}$ or $Mg^{2+}$ (Brown and Cook, 1981).

## Uptake of replacement, possibly toxic, metals

Replacement metals in general must enter the cell with difficulty in view of the absence of specific transport pathways. They may exert major effects inside the cell, perhaps out of proportion to their cellular concentration, but the limiting factor may well be the question of transport. Several possibilities are discussed in this section.

Microorganisms have selective transport systems for the uptake of the metals of known biological function, the multiplicity and complexity of which are sometimes remarkable, as in the examples of iron and potassium. In general, these systems are highly specific but systems of broader specificity are also present, which can also transport metal ions of no known function and which, if allowed to accumulate, would be toxic.

Clear examples of competition between essential and non-essential metal ions are provided by bacterial transport systems for monovalent and divalent cations. Thus, $Tl^+$ serves as a substrate for the potassium transport system in *Streptococcus lactis* (Kashket, 1979) and for the TrkA and Kdp potassium transport systems of *Escherichia coli* (Norris *et al.*, 1976; Damper *et al.*, 1979). In all cases uptake of $Tl^+$ is inhibited by $K^+$. It is clear that $Tl^+$ is a good probe for transport of potassium in both bacteria and yeasts (Norris *et al.*, 1976). However, much more is known about the transport of non-essential divalent ions.

## Transport of divalent cations via magnesium pathways

In view of the substantial requirement of bacteria for $Mg^{2+}$, it is not surprising that selective transport systems are available for its uptake. *Escherichia coli* has two distinct transport systems for $Mg^{2+}$ (Park *et al.*, 1976). One of these, System II, is specific for $Mg^{2+}$ but the other, System I, has high affinity for $Mg^{2+}$, with $K_m = 30 \mu M$, but also takes up $Mn^{2+}$, $Ni^{2+}$ and $Co^{2+}$ with lowered activity. The uptake of $Mn^{2+}$ or $Co^{2+}$ by System I is inhibitory or lethal, respectively, and leads to efflux of $Mg^{2+}$ from the cell. It is possible that accumulated $Mn^{2+}$ or $Co^{2+}$ displaces $Mg^{2+}$ from ribosomes, increasing intracellular 'free' magnesium concentration which is lost from the cell. It should be noted that highly specific transport systems for manganese have been found for a wide range of organisms (Silver and Jasper, 1977). Their high affinity for manganese allows the metal to be accumulated in the presence of much higher levels of calcium and magnesium. The *Escherichia coli* manganese transport system has a $K_m$ value of 0.2 $\mu M$ and is unaffected by a $10^5$-fold excess of $Mg^{2+}$.

## Transport of cadmium

Cadmium competes with both $Mn^{2+}$ and $Zn^{2+}$ transport systems. In *E.coli*, $Cd^{2+}$ appears to be a competitive inhibitor of a $Zn^{2+}$ transport system. Laddaga and Silver (1985) have shown exchange of cellular $^{109}Cd^{2+}$ with extracellular $Zn^{2+}$ and the competitive inhibition of uptake of $Cd^{2+}$ by $Zn^{2+}$, and have demonstrated the operation of a saturable transport system for $Cd^{2+}$. In contrast, $Cd^{2+}$ is taken up by a transport system for the active accumulation of $Mn^{2+}$ in Gram-positive bacteria such as *Staphylococcus aureus* (Tynecka *et al.*, 1981a,b; Perry and Silver, 1982), *Bacillus subtilis* (Laddaga *et al.*, 1985) and *Lactobacillus plantarum* (Archibald and Duong, 1984). The ability of the $Cd^{2+}$ ion to compete with both $Zn^{2+}$ and $Mn^{2+}$ for transport systems reflects the flexibility of cadmium in binding to either O- or N/S ligand domains. The *S.aureus* system has a high affinity for both $Cd^{2+}$ ($K_m = 5.4 \mu M$) and $Mn^{2+}$ ($K_m = 16 \mu M$). This chromosomally-encoded system is quite distinct from a second, plasmid-encoded system that serves to expel accumulated $Cd^{2+}$ (but not $Mn^{2+}$). Two plasmid loci confer different levels of resistance to $Cd^{2+}$. The better understood of these is the *cadA* system which confers high level resistance via a $Cd^{2+}/nH^+$ antiporter driven by the protonmotive force (Tynecka *et al.*, 1981a,b).

*Other pathways for the transport of non-essential metal ions*

Uptake of other metals into the cell is poorly understood, but is often associated with efflux of $K^+$. Copper and cadmium are rapidly accumulated by several fungi, following initial binding to the cell surface (Ross, 1975).

In some cases uptake of toxic metals into the cell is of biotechnological interest as a metal-removal process: thus, *Pseudomonas aeruginosa* takes up substantial quantities of uranium, which accumulates as intracellular deposits, and leads ultimately to the death of the organism. Uptake of uranium may involve the exploitation of a transport pathway for another metal ion of similar properties, but some shut-off mechanism fails to function and the uptake of uranium continues to toxic levels.

Non-essential metals may also be carried into the cell as complexes with a suitable ligand, which is probably negatively charged to give a neutral complex. Metal cations in the +III oxidation state may be taken into the cell as a complex with citrate or other ligands, as is known for $Fe^{3+}$. The involvement of higher oxidation state species in transport [e.g. Fe(III) as a complex with siderophores or other ligands and Co(III) as the $B_{12}$ coenzyme] enhances the selectivity of the process: a lower oxidation state cation would be more labile and less able to withstand competition from other metals. Manganese is taken up as Mn(II), but in this case it is not fulfilling a redox role and is probably not exposed to competition from the later transition metal.

Some heavy metals appear to enter cells directly, possibly through a lesion in the cell membrane resulting from the strong binding of the cation. Thallium(I) provides an example of this behaviour in its ability to penetrate into spheroplasts from *E.coli* even in the absence of an ionophore and the presence of the 'non-penetrant' sulphate anion (R.K.Poole, H.Butler and M.N.Hughes, unpublished work).

## Metal limitation of growth and function

Limitation of the supply of an essential metal to a microbial culture has two major applications in studies of metal−microbe interactions. First, a corollary to the conclusion that metal ions (with or without biological function) compete for biological sites, is that the outcome of such a competition can be influenced by limitation of the supply of one of these metals. Accordingly, concentrations of copper that were not toxic to *Legionella pneumophila* in a medium containing a range of metals had to be reduced 20-fold in a metal-deficient medium (Reeves *et al.*, 1981). Reduction in the concentration of an essential metal ion thus enhances the possibility that a foreign metal or a desired isotope can occupy a site in a metalloenzyme. Secondly, limitation of an essential metal may allow the normal function(s) of that metal to be defined, the limitation being manifest as a diminution or loss of a component such as metalloenzyme or, under extreme conditions, a decrease in growth rate and yield.

In principle, complex or defined media in either batch or continuous culture might be used in such investigations. In practice, the major experimental obstacle is the lowering of the concentration of the one metal in question to a level where biological

**Table 4.** Growth yields under metal-limited conditions. (Growth yields vary considerably with growth conditions.)

Approximate amount of cation (g) to give 100 g dry biomass

| | | | |
|---|---|---|---|
| $K^+$ | 1.7 | $Co^{2+}$ | 0.001 |
| $Mg^{2+}$ | 0.1−0.4 | $Cu^{n+}$ | 0.001 |
| $Ca^{2+}$ | 0.1 | $Zn^{2+}$ | 0.05 |
| $Mn^{2+}$ | 0.005 | $Mo^{n+}$ | 0.001 |
| $Fe^{n+}$ | 0.015 | | |

effects are demonstrable, and the poising of the culture in metal-limited growth for sufficient periods to allow experimental manipulation of other metal concentrations and the study of the effects brought about by these changes. For these reasons, the use of defined growth media in chemostat cultures is generally the first choice.

The methods involved in the setting up of metal-limited growth conditions vary from metal to metal, depending upon the levels of a metal required by the cell. The larger the requirement for a metal then the easier will it be to establish growth-limitation by that metal. It is not possible to define precisely the amounts of cation required by a particular organism as this will depend upon growth conditions and other factors. *Table 4* gives a broad indication of the amount of cation required to give 100 g of dry biomass when that cation is growth limiting (data taken from Pirt, 1975). The large requirement for potassium or magnesium results in the relatively easy achievement of metal-limited growth in these cases, often by sub-culturing the organism a number of times into metal-free growth medium. The high demand for $K^+$ and $Mg^{2+}$ means that the growth of the organism depletes the medium of these metal ions so that eventually their supply is growth-determining.

It is much more difficult to achieve these conditions for the transition metals, which are required in trace or ultra-trace amounts, and for elements such as sodium which are present as impurities in many components of growth media and cannot be satisfactorily removed. The amount of sodium adventitiously present will therefore usually be sufficient to supply the needs of the microorganism. Accordingly, in all these cases, metal-limitation will require not only the omission of the metal in question from the growth medium but also the rigorous extraction of residual metal introduced as a contaminant of other ingredients of the medium. Complex 'natural' components such as yeast extract, peptone and meat digests are rich sources of most metals required by microorganisms. Yeast extract, for example, contains all metals generally necessary for growth and has especially high levels of copper, iron, magnesium and zinc (Grant and Pramer, 1962). Indeed most media utilizing such ingredients rely entirely on them for metal supply. Thus the inability to control the provision of metals individually exacerbates the problem of extraction of the metal in question.

## Methods for the selective removal of metals from growth media

Several methods are available for lowering the concentration of metal ions in growth media. Earlier studies involved precipitation as hydroxides, phosphates, sulphides or carbonates (Hewitt, 1966; Lankford, 1973; Pirt, 1975) or adsorption on alumina (Ratledge and Chaudhary, 1971). These methods are not very specific, so it is now usual to use chelating cation exchange resins or solvent extraction with an organic solvent containing a chelating ligand that is reasonably selective for the metal to be removed. Chelex resin (BioRad) is a weak-acid cation exchanger (an iminodiacetic acid-substituted styrene derivative) which can be used to extract a variety of metal cations from media (see Bio-Rad Bulletins 2020 and 1224). It may be necessary to add other trace elements to the medium after extraction of the metal to ensure that growth is not limited by metals other than that desired. It is most essential that the actual concentration of the metal be determined in the metal-extracted medium. It is not sufficient merely to establish metal-limited growth. The behaviour of the microorganism may well depend upon the precise concentration of the metal which is growth limiting. This will determine the extent to which the replacement metal (and other metals inside the cell at normal physiological concentrations) will be able to compete and hence the effect upon the microorganism.

Metal-limitation studies, particularly with trace and ultra-trace metals, are technically demanding, requiring extreme attention to cleanliness and avoidance of metal contamination. Typically, water is doubly-distilled in an all-glass apparatus or distilled and then deionized (Hubbard *et al.*, 1986, 1989). Glassware and silicone components are brushed and autoclaved in 'Quadralene' (Fisons) and then rinsed repeatedly in distilled water. Glassware that cannot be autoclaved (for example, standard volumetric flasks) may be washed with chromic acid and rinsed repeatedly with distilled water.

*Preparation of iron-depleted medium.* Iron may be removed from growth medium by extracting with a solution of 8-hydroxyquinoline in chloroform (Waring and Werkman, 1942) or alternative solvents such as 1,2-dichloroethane (Partridge and Yates, 1982). The organic phase becomes black as the iron is extracted. The extraction is repeated as necessary and can give iron levels in the range 0.7 to 3.0 $\mu$g Fe l$^{-1}$. Needless to say, this is a very time-consuming and tedious operation. *Aerobacter aerogenes* grown in this medium proved to be deficient in catalase and in cytochromes (Waring and Werkman, 1944). Care should be taken to ensure that excess of 8-hydroxyquinoline is not present in the medium as this may be toxic.

Iron may also be extracted from medium by the use of Chelex 100. This method is reported to give iron levels of 0.3 $\mu$M (Arnold *et al.*, 1983).

*Preparation of copper-depleted medium.* Copper may be extracted from medium by the use of diethyldithiocarbamate plus solvent to give levels as low as 0.01 $\mu$M (Iwasaki *et al.*, 1980). Chelex resin may also be used (Schwab, 1973) as illustrated in the following typical procedure for copper limitation of *Paracoccus denitrificans* (Hubbard *et al.*, 1989). Copper was omitted from the trace elements solution normally added to the defined medium; the basal medium (at ten times the final concentration) was passed through a column containing Chelex resin. The final medium was made by dilution of the above solutions with double glass-distilled water. After assay for copper, the desired final concentration can be achieved, if appropriate, by addition of a concentrated solution of high purity (for example AristaR grade) $CuCl_2$.

### Growth vessels for use with metal-extracted media

Growth vessels must not have metallic components. Several all-glass or glass—plastic chemostat culture vessels are described in the literature. An effective but simple apparatus

(Hubbard *et al.*, 1986) is based on a water-jacketed Quickfit vessel (1 litre working volume) with appropriate ports for glass electrodes, and glass delivery tubes for medium and solutions of acid or alkali. Stirring was achieved using a glass paddle and shaft, journaled in a lubricated felt bearing (that is, non-metallic) and driven mechanically, but designs based upon magnetic stirring or hollow titanium drive shafts (which also transmit air to the culture) have been described (Ware *et al.*, 1970).

### Replacement of potassium by thallium

It has already been noted that $Tl^+$ is a good probe of potassium transport systems in bacteria and yeast (Norris *et al.*, 1976; Damper *et al.*, 1979; Kashket, 1979; Davidson and Knaff, 1982). It is not, of course, a specific replacement for $K^+$, and does serve as a substrate for some ammonium transport systems (Jayakumar *et al.*, 1985).

The potassium cation binds to oxygen donor ligands with low affinity, which makes the characterization of binding difficult. Furthermore, while potassium concentrations can be determined readily by atomic emission spectroscopy, the function of potassium is difficult to study. The isotope potassium-42 has an inconveniently short halflife of 12.36 h and the use of potassium-39 NMR techniques is not well developed. The cation $Rb^+$ is often used as a replacement marker for $K^+$ but is not of great value in studies on bacterial transport systems, which discriminate between these two cations (for example Benyoucef *et al.*, 1981). The best replacement for $K^+$ in terms of ionic radius is $Tl^+$ (*Table 3*), while thallium has a number of properties that may be exploited to characterize its environment, notably NMR. It also has an isotope, thallium-204, with a very convenient halflife of 3.9 years. On the other hand, thallium is well known to be a toxic element (Christie, 1985) and binds much more strongly to ligands than does potassium, often with formation constants $10^3$ times greater than those found for $K^+$. This behaviour is due to the greater electron density of thallium which results in mutual polarization effects between cation and ligand and stronger interaction; that is, thallium is a 'soft' cation. It should be recalled that thallium is in Group III of the Periodic Table and that the existence of the $Tl^+$ cation results from the inert pair effect. Thus, it is not surprising that $Tl^+$ has similarities to the Group IB element silver in addition to the IA element potassium. Silver is one of the most toxic metals to bacteria! The toxicity of thallium is likely to result from binding too strongly to potassium sites (behaving as a 'super potassium') and from binding to additional sites. The former possibility means that low concentrations of $Tl^+$ may well activate $K^+$-dependent processes more effectively than $K^+$. Both these effects were shown in the study of the replacement of potassium by thallium in yeast aldehyde dehydrogenase (Bostian *et al.*, 1982). In summary, while $Tl^+$ appears to be a good replacement for $K^+$, its behaviour deviates from that of thallium in an understandable way. It should also be noted that the solubilities of many simple thallium salts are much lower than those of potassium and precipitation problems may occur.

As noted earlier, potassium-limited growth can be achieved relatively easily by sub-culturing an organism under potassium-free conditions. For potassium-limited *E.coli*, the toxic effects of thallium were manifested in inhibition of growth, which was directly related to the concentration of $Tl^+$, and inhibition of respiration. In the latter case, 0.2 M $Tl^+$ exerted a greater effect on respiration than did 40 mM concentrations of the IA cations. The toxicity produced by $Tl^+$ could be alleviated by increasing the concentration of $K^+$ in the cases of *Bacillus megaterium* and *Saccharomyces cerevisiae*, suggesting the two cations were competing directly for binding sites. In the case of *E.coli*, added $K^+$ was much less effective in alleviating the toxicity of $Tl^+$.

## Iron limitation of growth: competition from $Ga^{3+}$ and $Al^{3+}$

Achieving iron-deficiency of growth, particularly in bacteria, is frustrated by the remarkable avidity and multiplicity of iron transport systems synthesized by micro-organisms to overcome the low solubility of iron compounds at pH values near 7 (Neilands, 1984). In *E.coli*, for example, at least five separate systems are known, the best understood of which involves enterobactin, a catecholate siderophore of remarkably high affinity for Fe(III) with $\log K = 52$ at pH 7.4 (Harris *et al.*, 1979). The synthesis of enterobactin is induced under iron-limiting conditions, whether natural or experimental. Similar difficulties have been experienced with yeast (for example, Clegg and Garland, 1971).

Iron limitation offers an alternative strategy for the study of iron-containing components of respiratory chains, although the experimental difficulties seem to have limited work in this area (for example, Rainnie and Bragg, 1973). We have satisfactorily lowered the iron content of a defined medium to concentrations ($0.3-1.5$ $\mu$M) where steady-state biomass concentration in continuous culture is significantly diminished and cultures are no longer succinate limited, as revealed by the appearance of unconsumed carbon source in the chemostat effluent. In batch culture, four successive sub-cultures from stationary phase cultures into fresh iron-extracted medium achieved a seven-fold reduction in the concentration of the final biomass (Hubbard *et al.*, 1986). Iron-limited chemostat-grown cells contained depressed levels of cytochromes and non-haem iron proteins and diminished respiration rates and respiration-driven proton-translocation quotients. Haemoprotein $b_{590}$ (a catalase and peroxidase) was reduced 20-fold in iron-limited cells.

Gallium(III), Fe(III) and Al(III) have similar ionic radii (*Table 5*) and each forms complexes with siderophores (Tait, 1975; Theil *et al.*, 1983), transferrin and citrate. Some formation constants are also shown in *Table 5*. The use of $Ga^{3+}$ as a probe for $Fe^{3+}$ in storage and transport systems has been noted. $Ga^{3+}$ is the only known cation whose kinetic and thermodynamic stability allows it significantly to displace $Fe^{3+}$ from its complexes with siderophores (Emery, 1986). A ten-fold excess of $Ga^{3+}$ displaces up to 30% of $Fe^{3+}$ from siderophores of the hydroxamic acid type. If this process is carried out under reducing conditions, the displaced $Fe^{3+}$ is reduced to $Fe^{2+}$ and the resulting shift in the equilibrium means that all of the $Fe^{3+}$ will be displaced from the siderophore. Not unexpectedly, $Ga^{3+}$ also binds citrate strongly and so may perturb high-affinity iron transport processes involving citrate. It is clear that $Ga^{3+}$ is a useful replacement for $Fe^{3+}$, particularly under iron-limited conditions.

The addition of $Ga^{3+}$ (9 $\mu$M) to iron-limited cultures of *E.coli* diminishes growth further (Hubbard *et al.*, 1986) suggesting interference with the enterobactin system as has also been described for *Neurospora crassa* (Winkelmann *et al.*, 1973), *Ustilago sphaerogena* (Emery, 1971; Emery and Hoffer, 1980) and *Paracoccus denitrificans* (Bergeron and Kline, 1984). In some experiments of this type, $Ga^{3+}$ was supplied as the citrate, which could result in stimulation of gallium uptake, while it should also be noted that Ga(III) citrate exists as a kinetically stable polymer. In *E.coli*, in batch

**Table 5.** Properties of $Fe^{3+}$, $Al^{3+}$ and $Ga^{3+}$.

|          | r(A)  | logK |             |         |
|----------|-------|------|-------------|---------|
|          |       | EDTA | Transferrin | Citrate |
| $Fe^{3+}$ | 0.64  | 25   | 22.7, 22.1  | 11.4    |
| $Al^{3+}$ | 0.50  | 16   | 12.9, 12.3  | 8.00    |
| $Ga^{3+}$ | 0.62  |      |             |         |

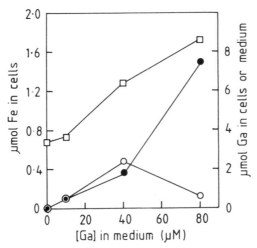

**Figure 1.** Effect of gallium(III) on the accumulation of iron and gallium by batch-grown cells of iron-sufficient *Escherichia coli*. Cells were grown in 100 ml cultures containing 20 $\mu$M [$Fe^{3+}$] and in the presence of the $Ga^{3+}$ concentrations shown; the medium and cells were then analysed for metal content. Shown are the iron content of cells from 100 ml of culture ($\square$) and the amount of gallium found in the cells ($\bigcirc$) or the medium ($\bullet$).

culture, uptake of $Ga^{3+}$ is greater in succinate-limited than in iron-limited or 'low iron' cultures. *Figure 1* shows the effect of gallium concentration on the accumulation of iron and gallium, with an initial iron concentration of 20 $\mu$M. Interestingly, the presence of $Ga^{3+}$ stimulates iron uptake and is itself taken up, indicating its potential value in exploration of the poorly understood low-affinity pathways for iron uptake. At 80 $\mu$M $Ga^{3+}$ more than 85% of the iron originally in the medium is taken up by the cells, so that intracellular iron is more than doubled.

The data in *Table 5* shows that $Al^{3+}$ does not compete well with $Fe^{3+}$, probably due to the small size of the aluminium cation which prevents effective binding of large ligands. We are currently examining the effect of $Al^{3+}$ on the growth of *E.coli* and the uptake of $Al^{3+}$ into the cell at different concentrations of $Fe^{3+}$, in order to model $Fe^{3+} - Al^{3+}$ competitions.

## Copper limitation of growth

Copper-limited growth is a potentially powerful approach to the study of copper function in a variety of copper-containing enzymes and proteins. Our interest in bacterial cytochrome oxidases, for example, has led us to investigate copper-limited growth of

*Paracoccus denitrificans* as an alternative to copper depletion by chemical means (for example, Weintraub *et al.*, 1982). This organism was the first bacterium from which cytochrome $aa_3$-type oxidase was characterized in detail and which proved to be structurally and functionally similar to the mitochondrial oxidase. In particular, the enzyme contains at least two and probably three moles of copper per mole of enzyme (Steffens *et al.*, 1987).

The copper concentration of a chemically defined medium could be reduced from 5 $\mu$M to an average level of 0.02 $\mu$M using Chelex resin. Even this level of copper did not greatly diminish growth yields but elicited striking changes in the content and spectral characteristics of cytochrome oxidase $aa_3$. In particular, changes in the near infrared optical spectrum suggest loss of $Cu_A$, the isolated copper centre (as distinct from $Cu_B$ which is anti-ferromagnetically coupled to cytochrome $a_3$). Interestingly, the oxidase appears to be functionally active, binding the oxygen molecule and terminating a phosphorylating respiratory chain, but displays unusually slow binding of carbon monoxide to haem $a_3$ in sub-zero temperature experiments (Hubbard *et al.*, 1988). This system should provide a unique opportunity to study the interaction between the various copper and haem centres of this enzyme.

Copper limitation of *Paracoccus* illustrates a number of potential and actual problems in such studies. First, metal deficiency is not necessarily reflected in changes of growth, rate or yield, if the metalloprotein or metalloenzyme requiring the metal is redundant or, if, as in the case of *Paracoccus* oxidase, only some of the copper atoms are essential for function. Likewise, when *Pseudomonas* AM1 is grown in batch culture in copper-deficient medium one copper atom is missing from the purified oxidase (Fukumori *et al.*, 1985). Second, working with extremely low levels of residual metal, close to the detection limits of some analytical techniques and to the level where unavoidable contamination by adventitious metal occurs, reproducing degrees of metal limitation can be problematic. Thus residual copper levels in Chelex-extracted medium range from 0.009 $\mu$M to 0.02 $\mu$M, but no attempt was made to study effects within this range.

## Modulation of silver toxicity by copper ions

Continuous culture also permits investigation of the competition by metals for binding sites by allowing the manipulation of the concentration of a metal at poised entering concentrations of other metals. This approach has recently been used to demonstrate the protection afforded by copper from the toxicity of silver in *E.coli* (Ghandour *et al.*, 1988).

Growth of *E.coli* in a chloride-free medium (to avoid precipitation of silver chloride) in either batch or continuous culture is inhibited by silver. The decrease in growth yield elicited by entering silver concentrations in the range 0.13−0.94 $\mu$M was alleviated by increases in the entering copper concentration, although at none of the copper concentrations used was the metal rate-limiting (or toxic). Thus toxicity is a function of the [Ag]/[Cu] ratio rather than the silver concentration *per se*; growth yield is unaffected at [Ag]/[Cu] below about 0.3, but declines to 20% of control values (that is, without added silver) at [Ag]/[Cu] $\geq 0.5$ (Ghandour *et al.*, 1988). These results suggest that silver and copper compete for sites at which $Ag^+$ exerts its toxic effects and that high copper concentrations protect cells by virtue of such competition. The alternative explanation is that silver competes for sites via which copper is transported or at which copper functions so that the presence of silver leads in effect to copper limitation. Indeed recent experiments have demonstrated, in the absence of silver, copper-limitation of growth of *E.coli* (J.A.Hubbard, R.K.Poole and M.N.Hughes, unpublished). The lesions have yet to be identified but might be involved in cytochrome oxidase *o* (Kita *et al.*, 1984) or another copper enzyme, most of which are involved in biological electron transfer of oxygen utilization (Cass and Hill, 1980).

## References

Archibald,F.S. and Duong,M.-N. (1984) Manganese acquisition by *Lactobacillus plantarum*. *Journal of Bacteriology* **158**, 1−8.

Arnold,W.N., Evans,B.J. and Denniston,M.L. (1983) *Journal of General Microbiology* **129**, 2351−2358.

Benyoucef,M., Rigaud,J.L. and Leblanc,G. (1981) The electrochemical proton gradient in mycoplasma cells. *European Journal of Biochemistry* **113**, 491−498.

Bergeron,R.J. and Kline,S.J. (1984) 300 MHz ${}^1$NMR study of parabactin and its gallium(III) chelate. *Journal of the American Chemical Society* **106**, 3089−3098.

Bostian,K., Betts,G.F., Man,W.K. and Hughes,M.N. (1982) Multiple binding of thallium and rubidium to potassium-activated yeast aldehyde dehydrogenase. Influences on tertiary structure, stability and catalytic activity. *Biochemical Journal* **207**, 73−80.

Brown,D.A. and Cook,R.A. (1981) Role of metal cofactors in enzyme regulation. Differences in the regulatory properties of the *Escherichia coli* nicotinamide adenine dinucleotide phosphate specific malic enzyme, depending on whether magnesium ion or manganese ion serves as divalent cation. *Biochemistry* **20**, 2503−2512.

Cass,A.E.G. and Hill,H.A.O. (1980) Copper proteins and copper enzymes. In *Biological Roles of Copper*, (CIBA Foundation Symposium 79). Excerpta Medica, Amsterdam, pp. 71−91.

Christie,A. (1985) *The Pale Horse*. Dodd, London.

Clegg,R.A. and Garland,P.B. (1971) Non-haem iron and the dissociation of piericidin A sensitivity from *Torulopsis utilis*. *Biochemical Journal* **124**, 135−154.

Cohn,M. (1970) Magnetic resonance studies of enzyme−substrate complexes with paramagnetic probes as illustrated by creatine kinase. *Quarterly Review of Biophysics* **3**, 61−82.

Damper,P.D., Epstein,W., Rosen,B.P. and Sorensen,E.N. (1979) Thallous ion is accumulated by potassium transport systems in *Escherichia coli*. *Biochemistry* **18**, 4165−4169.

Davidson,V.L. and Knaff,D.B. (1982) ATP-dependent K${}^+$ uptake by a photosynthetic purple sulfur bacterium. *Archives of Biochemistry and Biophysics* **213**, 358−362.

Edwardson,J.A., Klinowski,J., Oakley,A.E., Perry,R.H. and Candy,J.M. (1986) Aluminosilicate and the ageing brain: implications for the pathogenesis of Alzheimer's disease. In *Silicon Biochemistry* (Ciba Foundation Symposium 121). Wiley, Chichester, pp. 160−179.

Emery,T. (1971) Role of ferrichrome as a ferric ionophore in *Ustilago sphaerogena*. *Biochemistry* **10**, 1483−1488.

Emery,T. (1986) Exchange of iron by gallium in siderophores. *Biochemistry* **25**, 4629−4633.

Emery,T. and Hoffer,P.B. (1980) Siderophore-mediated mechanism of gallium uptake demonstrated in the microorganism *Ustilago sphaerogena*. *Journal of Nuclear Medicine* **21**, 935−939.

Foye,W.O. (1977) Antimicrobial activities of mineral elements. In *Microorganisms and Minerals* (ed. E.D.Weinberg), Marcel Dekker, New York, pp. 387−420.

Friberg,L., Nordberg,G.F. and Vouk,V.B. (1979) Silver. In *Handbook on the Toxicology of Metals*, Elsevier/North Holland Biomedical Press, Amsterdam, pp. 579−585.

Fukumori,Y., Nakayama,K. and Yamanaka,T. (1985) One of two copper atoms is not necessary for the cytochrome *c* oxidase activity of *Pseudomonas* AM1 cytochrome *aa₃*. *Journal of Biochemistry* **98**, 1719−1722.

Ghandour,W., Hubbard,J.A., Deistung,J., Hughes,M.N. and Poole,R.K. (1988) The uptake of silver ions by *Escherichia coli* K12: toxic effects and interaction with copper ions. *Applied Microbiology and Biotechnology* **28**, 559−565.

Grant,C.L. and Pramer,D. (1962) Minor element composition of yeast extract. *Journal of Bacteriology* **84**, 869−870.

Harris,W.R., Carrano,C.J. and Raymond,K.N. (1979) Spectrophotometric determination of the proton-dependent stability constant of ferric enterobactin. *Journal of the American Chemical Society* **101**, 2213−2214.

ALLEGHENY COLLEGE LIBRARY

15

Hayes,R.L. (1978) The medical use of gallium radionuclides. A brief history with some comments. *Seminar on Nuclear Medicine* **8**, 183−190.

Hayes,R.L. and Hubner,K.F. (1983) Basis for the clinical use of gallium and iridium radionuclides. In *Metal Ions in Biological Systems (ed. H.Sigel)* **16**, Marcel Dekker, New York, pp. 279−315.

Hewitt,E.J. (1966) *Sand and Water Culture Methods used in the Study of Plant Nutrition.* 139−143 and 455. Commonwealth Agricultural Bureau, Farnham, Bucks.

Hubbard,J.A.M., Lewandowska,K.B., Hughes,M.N. and Poole,R.K. (1986) Effects of iron-limitation of *Escherichia coli* on growth, the respiratory chains and gallium uptake. *Archives of Microbiology* **146**, 80−86.

Hubbard,J.A., Hughes,M.N. and Poole,R.K. (1989) Effects of copper concentration in continuous culture on the $aa_3$-type cytochrome oxidase and respiratory chains of *Paracoccus denitrificans*. *Archives of Microbiology* (in press).

Hughes,M.N. (1981) *The Inorganic Chemistry of Biological Processes*, 2nd edn., Wiley, Chichester.

Hughes,M.N. (1987) Coordination compounds in biology. In *Comprehensive Coordination Chemistry* (eds G.Wilkinson, R.D.Gillard and J.A.McCleverty) Pergamon Press, Oxford, **6**, pp. 545−754.

Iwasaki,H., Saigo,T. and Matsubara,T. (1980) Copper as a controlling factor of anaerobic growth under $N_2O$ and biosynthesis of $N_2O$ reductase in denitrifying bacteria. *Plant and Cell Physiology* **21**, 1573−1584.

Jayakumar,A., Epstein,W. and Barnes,E.M. (1985) Characterisation of ammonium (methylammonium)/potassium antiport in *Escherichia coli*. *Journal of Biological Chemistry* **260**, 7528−7532.

Kashket,E.R. (1979) Active transport of thallous ion by *Streptococcus lactis*. *Journal of Biological Chemistry* **254**, 8129−8131.

Kita,K., Konishi,K. and Anraku,Y. (1984) Terminal oxidases of *Escherichia coli* aerobic respiratory chain. *Journal of Biological Chemistry* **259**, 3368−3374.

Laddaga,R.A. and Silver,S. (1985) Cadmium uptake in *Escherichia coli* K12. *Journal of Bacteriology* **162**, 1100−1105.

Laddaga,R.A., Bessen,R. and Silver,S. (1985) Cadmium-restricted mutant of *Bacillus subtilis* 168 with reduced cadmium transport. *Journal of Bacteriology* **162**, 1106−1110.

Lankford,C.E. (1973) Bacterial assimilation of iron. *CRC Critical Reviews in Microbiology* **2**, 273−331.

MacDonald,T.L. and Martin,R.B. (1988) Aluminium ion in biological systems. *Trends in Biochemical Sciences* **13**, 15−19.

Martin,R.B. (1986) The chemistry of aluminium as related to biology and medicine. *Clinical Chemistry* **32**, 1797−1806.

Mildvan,A.S. (1977) Magnetic resonance studies of the conformations of enzyme-bound substrates. *Accounts of Chemical Research* **10**, 246−252.

Mildvan,A.S. (1979) The role of metals in enzyme-catalysed substitutions at each of the phosphorus atoms of adenosine triphosphate. *Advances in Enzymology* **49**, 103−126.

Neilands,J.B. (1984) Siderophores of bacteria and fungi. *Microbiological Science* **1**, 9−14.

Norris,P.R., Man,W.K., Hughes,M.N. and Kelly,D.P. (1976) Toxicity and accumulation of thallium in bacteria and yeast. *Archives of Microbiology* **110**, 279−286.

Park,M.H., Wong,B.B. and Lusk,J.E. (1976) Mutants in three genes affecting transport of magnesium in *Escherichia coli*: genetics and physiology. *Journal of Bacteriology* **126**, 1096−1103.

Partridge,C.D.P. and Yates,M.G. (1982) Effect of chelating agents on hydrogenase in *Azotobacter chroococcum*. *Biochemical Journal* **204**, 339−344.

Perry,R.D. and Silver,S. (1982) Cadmium and magnesium transport of *Staphylococcus aureus* membrane vesicles. *Journal of Bacteriology* **150**, 973−976.

Pirt,S.J. (1975) *Principles of Microbe and Cell Cultivation*. Blackwell, Oxford.

Rainnie,D.J. and Bragg,P.D. (1973) The effect of iron deficiency on respiration and energy-coupling in *Escherichia coli*. *Journal of General Microbiology* **77**, 339−349.

Ratledge,C. and Chaudhary,M.A. (1971) Accumulation of iron-binding phenolic acids by Actinomycetales and other organisms related to mycobacteria. *Journal of General Microbiology* **66**, 71−78.

Reeves,M.W., Pine,L., Hutner,S.H., George,J.R. and Harrell,W.E. (1981) Metal requirements of *Legionella pneumophila*. *Journal of Clinical Microbiology* **13**, 688−695.

Ross,I.S. (1975) Some effects of heavy metals on fungal cells. *Transactions of the British Mycological Society* **64**, 175−193.

Schwab,A.J. (1973) Mitochondrial protein synthesis and cyanide-resistant respiration in copper-depleted cytochrome oxidase-deficient *Neurospora crassa*. *FEBS Letters* **35**, 63−66.

Sillen,L.G. and Martell,A.E. (1971) *Stability Constants*. Special Publication No. 25, The Chemical Society, London.

Silver,S. and Jasper,P. (1977) Manganese transport in microorganisms. In *Microorganisms and Minerals* (ed. E.D.Weinberg), Dekker, New York, pp. 105−150.

Steffens,G.C.M., Biewald,R. and Buse,G. (1987) Cytochrome *c* is a three-copper, two haem-A protein. *European Journal of Biochemistry* **164**, 295−300.

Summers,M.F. (1988) [113]Cd NMR spectroscopy of coordination compounds and proteins. *Coordination Chemistry Reviews* **86**, 43−134.

Tait,H.G. (1975) The identification and biosynthesis of siderochromes formed by *Micrococcus denitrificans*. *Biochemical Journal* **146**, 191−204.

Theil,E.C., Eichhorn,G.L. and Marzilli,L.G. (1983) Iron-binding proteins without cofactors or sulfur clusters. In *Advances in Inorganic Biochemistry* **55**, Elsevier, New York.

Tynecka,Z., Gos,Z. and Zajac,J. (1981a) Reduced cadmium transport determined by a resistance plasmid in *Staphylococcus aureus*. *Journal of Bacteriology* **147**, 305−312.

Tynecka,Z., Gos,Z. and Zajac,J. (1981b) Energy-dependent efflux of cadmium coded by a plasmid resistance determinant in *Staphylococcus aureus*. *Journal of Bacteriology* **147**, 313−319.

Vallabhajosula,S.R., Harwig,J.F., Siemsen,J.K. and Wolf,W. (1980) Behaviour of gallium-67 in the blood: the role of transferrin. *Journal of Nuclear Medicine* **21**, 650−656.

Viola,R.E., Morrison,J.F. and Cleland,W.W. (1980) Interaction of metal(III)-adenosine 5′-triphosphate complexes with yeast hexokinase. *Biochemistry* **19**, 3131−3137.

Ware,G.C., Thompson,A.S. and Light,P.A. (1970) A non-metallic chemostat for micro-organisms requiring trace metal concentrations. *Laboratory Practice* **19**, 181−182.

Waring,W.S. and Werkman,C.H. (1944) Iron deficiency in bacterial metabolism. *Archives of Biochemistry* **4**, 75−87.

Weintraub,S.T., Muhoberac,B.B. and Wharton,D.C (1982) The effects of copper depletion on structural aspects of cytochrome *c* oxidase. *Journal of Biological Chemistry* **257**, 4940−4946.

Welch,M.J. and Moerlein,S. (1980) Radio-labeled compounds of biomedical interest containing radioisotopes of gallium and indium. *ACS Symposium Series* **140**, 122−140.

Williams,R.J.P. (1981) Physico-chemical aspects of inorganic element transfer through membranes. *Philosophical Transactions of the Royal Society of London* **B294**, 57−74.

Winkelmann,G., Barnekow,A., Ilgner,D. and Zahner,H. (1973) Metabolic products of microorganisms. 120. Uptake of iron by *Neurospora crassa* II. Regulation of the biosynthesis of sideramines and inhibition of iron transport by metal analogues of coprogen. *Archives of Microbiology* **92**, 285−300.

Womack,F.C. and Colowick,S.P. (1979) Proton-dependent inhibition of yeast and brain hexokinase by aluminium in ATP preparations. *Proceedings of the National Academy of Sciences USA* **76**, 5080−5084.

CHAPTER 2

# Heavy metal and radionuclide accumulation and toxicity in fungi and yeasts

G.M.GADD and C.WHITE

*Department of Biological Sciences, University of Dundee, Dundee DD1 4HN, UK*

## Introduction

Approximately 65 elements exhibit metallic properties and most of these may be arbitrarily termed heavy metals (Duxbury, 1985). Although many are essential components of biological systems, all are potentially toxic. The major physiological functions and mechanisms of toxicity both result from properties of these elements which include electrical charge, coordinating abilities and the possession of multiple valency states. These govern many aspects of the metabolism of these elements, particularly intracellular sequestration by binding molecules or in the vacuole. Toxic effects include the blocking of functional groups of biologically important molecules, e.g. enzymes, polynucleotides or transport systems for essential nutrients and ions, the displacement and/or substitution of essential metal ions from biomolecules and functional cellular units, conformational modification, denaturation and inactivation of enzymes and disruption of cell and organellar membrane integrity (Ochiai, 1987). Thus, toxic symptoms often vary widely between different organisms and for different metals. However, as many metals are essential for growth and metabolism, e.g. copper, zinc, manganese, iron and cobalt, fungi (as well as other microorganisms) possess mechanisms for their intracellular accumulation from the external environment at low concentrations. Many other metals and radionuclides, e.g. lead, tin, cadmium, aluminium, mercury, uranium and thorium, have no essential biological functions but can still be accumulated.

A prerequisite for toxic interactions is contact between the active metal species and cellular components and transport systems of varying specificity may be utilized (Gadd, 1986a, 1989a). Non-essential metals are generally of low abundance in the biosphere and should therefore not compete with specific transport systems for essential metals (Wood and Wang, 1983). However, due to industrial activities and deliberate and accidental discharges, microorganisms are increasingly exposed to potentially toxic conditions and may need to respond using a variety of strategies that ensure reproduction and survival. Thus, the characteristics and mechanism(s) of metal uptake are of central importance to any study of metal−microbe interactions and a relationship between uptake and toxicity is frequently observed, many resistance mechanisms relying on decreased uptake or impermeability (Gadd, 1989a). These aspects of metal−fungal interactions will be dealt with in this chapter as well as their application in novel biotechnological processes for the removal and/or recovery of heavy metals and radionuclides from contaminated aqueous effluents and process streams (Gadd, 1986b, 1989b).

## Heavy metal toxicity and resistance

### Inorganic metal compounds

In view of the wide spectrum of potentially toxic interactions between metals and microbial cells, almost every aspect of microbial metabolism and activity can be affected including respiration, ribosome synthesis and activity and membrane transport (Duxbury, 1985; Ochiai, 1987; Gadd, 1989a). The cell membrane is the obvious first site of action for any toxic agent and extensive membrane damage can be caused by heavy metals resulting in loss of cellular solutes and permeabilization of the cell to external materials (Passow and Rothstein, 1960; Norris and Kelly, 1977; Kuypers and Roomans, 1979; White and Gadd, 1987a,b). However, this is not universal as *Saccharomyces cerevisiae* lost $K^+$ in response to $Zn^{2+}$ at toxic concentrations while *Sporobolomyces roseus* did not (Mowll and Gadd, 1983). Copper caused extensive bursting of protoplasts of a copper-sensitive yeast strain but not of protoplasts from a resistant strain (Gadd *et al.*, 1984a).

Indirect mechanisms of heavy metal toxicity may involve free radicals. These reactive species are very damaging to cells as they can take part in chain reactions which involve the breakdown of biological macromolecules. Consequently aerobic organisms possess protective enzymes such as superoxide dismutase to eliminate those produced by normal metabolism. A major target of free radicals in cells are membranes, where they initiate lipid peroxidation in a chain reaction whereby the alkyl chains of lipids are converted to peroxyalkyl radicals and fatty acid hydroperoxides (Mehlhorn, 1986). Lipid-soluble complexes of transition elements such as Fe(II) and Cu(I) may undergo the Fenton reaction (1,2) with the latter and accelerate this process (McCord and Day, 1978).

$$Fe^{2+} + H_2O_2 \rightarrow Fe^{3+} + OH^- + OH^{\cdot} \tag{1}$$

$$O_2^- + Fe^{3+} \rightarrow O_2 + Fe^{2+} \tag{2}$$

Complexes and free ions of these cations may also undergo this reaction in aqueous solution. In animal systems, metal ions such as $Hg^{2+}$, $Cd^{2+}$, $Pb^{2+}$ and $Ag^+$ induce free radical toxicity as a result of their reactions with thiols or enzymes which normally protect against these reactive species (Mehlhorn, 1986).

### Organometallic compounds

Organometallic compounds have an increasing significance as a result of their use in the chemicals and petroleum industries and as biocides. Organometals are generally more toxic towards fungi than free metal ions and the toxicity of organometal compounds varies both with the number and with the identity of the organic groups (Blunden *et al.*, 1984; Cooney and Wuertz, 1989).

The major effects of organotins and organoleads are disruption of mitochondrial membranes and action as $Cl^-/OH^-$ ionophores depolarizing electrochemical gradients and consequently interfering with energy conservation (Blunden *et al.*, 1984; Cooney and Wuertz, 1989). Organometals may also damage membranes by the production of free radicals since the carbon−metal bond readily reacts with any available radicals to produce peroxyalkyl radicals which can result in lipid peroxidation (Mehlhorn, 1986). As well as the mitochondrial membrane, organometallic compounds may also exert a disruptive effective on the cell membrane and can cause a loss of solutes from yeast cells (Cooney *et al.*, 1989).

Fungal resistance towards organometallic compounds shows many of the same mechanisms as that towards inorganic metal compounds. Phenylmercuric acetate resistance by *Pyrenophora avenae* resulted from binding of the compound by extra-cellular pigments or intracelluar molecules (Greenaway, 1972). However, because toxicity depends on the nature and extent of substitution, transformation of organic and inorganic metal compounds to more volatile or less toxic species may be an important resistance mechanism. For example, methylation of $Hg^{2+}$ can be mediated by bacteria, fungi or yeasts. Although methylated species are more toxic, they are volatile and lost from the medium (Cooney and Wuertz, 1989). Organotins can be dealkylated by a variety of microbes to less toxic forms (Cooney and Wuertz, 1989).

### Accumulation of heavy metals and radionuclides

Fungi and yeasts can accumulate heavy metals and radionuclides even from dilute external concentrations. There are considerable differences in the uptake mechanisms employed depending upon whether organisms are living or dead and, if living, whether growth or cell differentiation takes place (Gadd, 1989b). Metal uptake by fungi and yeasts may be biphasic. The first phase is metabolism-independent and therefore can also occur in dead and non-metabolizing cells, the cell-wall being the major site of uptake. Metabolism-dependent uptake consists of transport across the plasma membrane into the cells. These phases of uptake may not be seen in all fungi or with all metals and radionuclides. In growing cultures, metabolism-dependent and -independent processes may be affected by the excretion of substances that complex or precipitate metals, changes in the chemical nature of the growth medium and cell differentiation (Gadd, 1989c).

### *Extracellular precipitation, binding and complexation*

Many extracellular fungal products can remove heavy metals from solution. Citric acid is an efficient chelating agent while oxalic acid production can result in precipitation of insoluble metal oxalates around cell-walls and in the external medium (Murphy and Levy, 1983). Hydrogen sulphide production by yeasts can result in metal precipitation in and around cell-walls (Minney and Quirk, 1985). Copper-grown strains of *S. cerevisiae* appear dark brown in colour because of the formation of copper sulphide.

Precipitation within or on cell-walls may be particularly evident with radionuclides such as uranium and thorium (*Figure 1*). In *S. cerevisiae*, uranium was deposited as a layer of needle-like fibrils on cell-walls reaching up to 50% of the biomass dry weight. That such a large amount was bound by the cells implied that additional uranium had 'crystallized' on already bound molecules (Strandberg *et al.*, 1981). Similar processes may occur in other organisms and this explains why radionuclide uptake may be a relatively slow process with some biomass types (Gadd *et al.*, 1988; Gadd, 1989b).

21

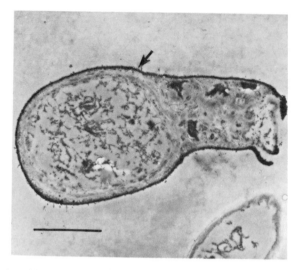

**Figure 1.** The location of thorium taken up by non-living *Rhizopus arrhizus* biomass in the cell-wall (arrowed). *R.arrhizus* was cultured in a glucose/salts medium (White and Gadd, 1988) for 72 h and exposed to thorium as a 0.5 mM solution in 1.0 M $HNO_3$ for 1 h at 25°C. Biomass was separated by filtration, washed and then fixed in 2% (v/v) glutaraldehyde in Hepes buffer, pH 6.8, at room temperature for 2 h. Dehydration was in an ascending ethanol series followed by an ethanol/propylene oxide series. Finally, cells were embedded in Epon 812 (T.J.Beveridge, personal communication). Sections were cut and examined without staining. The bar marker indicates 2.0 $\mu$m.

Many fungi and yeasts release high-affinity iron-binding molecules called siderophores. The excretion of these compounds can be markedly stimulated by iron deficiency and siderophores may also bind other metals, for instance gallium and nickel (Adjimani and Emery, 1987; Gadd, 1989b). *Debaryomyces hansenii* excreted riboflavin, or an analogous compound, under conditions of iron depletion or in the presence of copper, cobalt or zinc and this compound was also capable of binding $Fe^{3+}$ (Gadd and Edwards, 1986). It is conceivable that, in some circumstances, a secondary role of siderophore excretion may be some protection from metal toxicity.

## Metabolism-independent accumulation of heavy metals and radionuclides

Although simplistic in a biological context, adsorption is a term frequently used to describe metabolism-independent accumulation of heavy metals and radionuclides to fungal biomass. Adsorption may be defined as the accumulation of substances at a surface or interface whereas absorption is where the atoms or molecules of one phase almost uniformly penetrate into those of another phase forming a 'solution' with the second phase. The term 'sorption' can include both adsorption and absorption and describes the movement of a component from one phase to be accumulated in another solid phase (see Gadd, 1989b). Thus 'biosorption' is now frequently used to describe these physico-chemical reactions that occur between metals and cells (Shumate and Strandberg, 1985; Gadd, 1989b,c).

Metabolism-independent biosorption is generally rapid and virtually unaffected over modest ranges of temperature, e.g. 4−30°C (Norris and Kelly, 1977; de Rome and Gadd, 1987). A variety of ligands may be involved in metal binding, including carboxyl, amine, hydroxyl, phosphate and sulphydryl groups, and considerable differences in biosorption capacities are found between different species, strains and cell types (Gadd, 1986a,b, 1989b). Biosorption in *Rhizopus arrhizus* was related to ionic radius for a variety of metal ions, e.g. $Cu^{2+}$, $Zn^{2+}$, $Cd^{2+}$, $Hg^{2+}$, $UO_2^{2+}$, $Ag^+$ but not $Cr^{3+}$. Alkali metal cations, $Na^+$, $K^+$, $Rb^+$ and $Cs^+$ were not taken up (Tobin *et al.*, 1984). Low external pH frequently decreases the rate and extent of biosorption and other anions and cations can also affect biosorption by precipitation, e.g. phosphates, hydroxides or by competition for binding sites, e.g. $Mg^{2+}$, $Ca^{2+}$, $Zn^{2+}$, $Cu^{2+}$ and $Mn^{2+}$ (Gadd, 1986a, 1989b).

Metal-binding sites in cell-walls may not be involved in subsequent influx; surface biosorption may be reduced, or even absent, without any effect on rates of intracellular uptake (Borst-Pauwels, 1981; Mohan *et al.*, 1984).

Living or dead fungal biomass is capable of biosorption which, in general, is strongly dependent on biomass density. At a given equilibrium metal concentration, the specific uptake of heavy metals by fungal biomass, when expressed on a dry weight or per cell basis, for example, is greater at low cell densities than at high ones (de Rome and Gadd, 1987). Certain treatments may improve fungal biosorption. Powdering of dried biomass exposes additional binding sites (Tobin *et al.*, 1984) as does detergent treatment which increases cell permeability (Ross and Townsley, 1986; Gadd *et al.*, 1988). As well as dissolved metal forms, particulate material can also be bound by fungal biomass (see Chapter 8).

*Biosorption by fungal wall constituents.* Compounds derived from or produced by fungi can act as efficient biosorptive agents. A wide variety of such materials are found in fungal cell-walls, including glucans, mannans, melanins, chitin and chitosan. Chitin phosphate and chitosan phosphate removed greater amounts of $UO_2^{2+}$ from solution than other divalent metal cations and were more efficient than non-phosphorylated chitin and chitosan (Sakaguchi and Nakajima, 1982). Glucans with amino acid or sugar acid groups exhibit enhanced chelating abilities and rapid binding of metal ions (Muzzarrelli *et al.*, 1986). Fungal phenolic polymers and melanins contain phenolic units, peptides, carbohydrates, aliphatic hydrocarbons and fatty acids with oxygen-containing groups, e.g. carboxyl, phenolic, alcoholic hydroxyl, carbonyl and methoxyl, involved in metal binding (Saiz-Jimenez and Shafizadeh, 1984). Extracellular melanin from pigmented strains of *Aureobasidium pullulans* and *Cladosporium resinae* had a greater biosorption capacity for $Cu^{2+}$ than intact biomass, and enhanced $Cu^{2+}$ uptake was exhibited by pigmented biomass as compared with that from hyaline strains (Gadd and Mowll, 1985; Gadd and de Rome, 1988). Waste fungal biomass can arise in great quantities from several industrial fermentations and may provide a cheap source of potential biosorptive agents for industrial purposes. The role of extracellular polymers and polysaccharides in metal biosorption has received little attention in fungi.

*Biosorption of actinide elements and radionuclides.* Actinides and other radionuclides are of considerable industrial and environmental importance although most have no essential physiological role. Uptake has been examined in a wide range of fungi and consists predominantly of surface biosorption although the characteristics of uptake vary. Uranium and thorium uptake by *R. arrhizus* fitted a Freundlich isotherm at moderate pH values (Tsezos and Volesky, 1981) but it was not possible to relate uptake to any simple isotherm in several other studies. This is to be expected, as adsorption of solutes by solids is affected by a variety of factors such as diffusion and heterogeneity of the surface, which complicate the isotherm (Oscik, 1982), and the proportion of different actinide species present in solution which is strongly pH-dependent (Hunt, 1963). A more complex three-stage kinetic treatment yielded a good fit for kinetics and equilibria of $UO_2^{2+}$ adsorption by *S.cerevisiae* (Weidemann *et al.*, 1981). The morphology of filamentous fungi requires that an adequate theoretical treatment of these would allow for diffusion equilibria as well as the differentiation of the mycelial surface. The main site of uptake of actinides by fungi and yeasts appears to be the cell-wall (Weidemann *et al.*, 1981; Tsezos and Volesky, 1982a,b) although treatments that permeabilize the cell membrane such as alkali carbonates (Galun *et al.*, 1983a,b) or detergents (White and Gadd, 1988) can also increase uptake, indicating that sites within the cell are also capable of binding actinide elements. The mechanism of biosorption appears to vary between different actinide elements. Both adsorption and precipitation of hydrolysis products were involved for uranium (Tsezos and Volesky, 1982a) while the main mechanism of thorium biosorption was coordination with cell-wall nitrogen (Tsezos and Volesky, 1982b). The main cell-wall component responsible for biosorption in both of these instances was chitin (Tsezos, 1983).

### Metabolism-dependent accumulation

Energy-dependent metal uptake has been demonstrated for most physiologically important metal ions in many species of fungi and yeasts (*Table 1*; Borst-Pauwels, 1981; Gadd, 1986a,b, 1989b). Adsorption was the main mechanism of metal cation uptake by some filamentous fungi (Duddridge and Wainwright, 1980; Townsley and Ross, 1986) but the adsorptive capacity of the mycelial walls may be large compared with the available metal in solution and the fungal capacity for energy-dependent uptake, necessitating special methods such as the use of isolated protoplasts in order to demonstrate it (Theuvenet and Bindels, 1980; Gadd and White, 1985). Wall binding may also be very high in resting structures, as in *A.pullulans* chlamydospores which have no apparent metal transport. However, isolated chlamydospore protoplasts did show energy-dependent metal uptake but utilized a different transport system to other cell types (Gadd *et al.*, 1987).

*Energization of heavy metal transport.* Heavy metal transport is sensitive to conditions that inhibit cell energy metabolism such as the absence of substrate, anaerobiosis, incubation at low temperatures or the presence of respiratory inhibitors such as cyanide or azide (Paton and Budd, 1972; Norris and Kelly, 1977; Borst-Pauwels, 1981). Non-metabolizable glucose analogues such as 2-deoxy-D-glucose do not substitute for glucose (Borst-Pauwels, 1981).

*ATPase involvement.* Heavy metal transport by fungi and yeasts is dependent on ATPase activity (Borst-Pauwels, 1981) and ATPase inhibitors inhibit metal uptake (Failla *et al.*, 1976; Borst-Pauwels, 1981; White and Gadd, 1987a). Thus, diethylstilbestrol (DES) is an effective inhibitor *in vitro* and *in vivo* of the fungal membrane ATPase and its proton-pumping activity (Bowman *et al.*, 1978; Serrano, 1985) DES is also an inhibitor of fungal transport of divalent cations including heavy metals (Borst-Pauwels, 1981; White and Gadd, 1987a; Budd, 1988). Heavy metal uptake is also inhibited by other substances that depolarize the cell membrane such as protonophoric uncoupling agents. Uncouplers such as carbonylcyanide-*m*-chlorophenylhydrazone (CCCP) (Budd, 1988) or dinitrophenol (DNP) (Gadd and Mowll, 1985; Parkin and Ross, 1986; Hockertz *et al.*, 1987; White and Gadd, 1987a) also inhibit heavy metal uptake by fungi. However, these uncouplers also inhibit respiration-dependent ATP synthesis so that it is likely that some of the inhibitory effect is due to this. The $K^+$ ionophore, nigericin, stimulated $Cu^{2+}$ uptake by *A.pullulans*, presumably due to hyperpolarization resulting from increased efflux of internal $K^+$ (Gadd and Mowll, 1985). Conversely, a high external $K^+$ concentration inhibits divalent cation transport, apparently by depolarizing the cell membrane (Roomans *et al.*, 1979; Borst-Pauwels, 1981). Consequently, a direct role for the membrane ATPase in divalent cation transport has been discounted (Borst-Pauwels, 1981). The fungal cell membrane $H^+$-ATPase exports $H^+$ across the cell membrane energized by ATP hydrolysis (Slayman, 1970; Slayman *et al.*, 1973; Goffeau and Slayman, 1981). The protonmotive force thus generated consists of two components, an inside-alkaline pH gradient and an inside-negative membrane potential, both of which are possible energy sources for transport of solutes into the cell (Cooper, 1982). $K^+$ uptake is coupled to the $H^+$-ATPase via the membrane potential (Peña, 1975; Ramos *et al.*, 1985). Both $K^+$ and $Mg^{2+}$ uptake also stimulate $H^+$ efflux (Borst-Pauwels, 1981) but this does not occur with heavy metals and they can inhibit $H^+$ efflux from yeast cells *in vivo* (White and Gadd, 1987a,b). Thus the role of the $H^+$-ATPase in heavy metal uptake appears to be one of energizing the cell membrane rather than being directly coupled to heavy metal uptake.

*The role of the $K^+$ gradient and $K^+$ efflux.* $K^+$ is actively concentrated in fungal cells and mycelium, $K^+$ uptake being coupled to ATPase-dependent $H^+$ efflux via the membrane potential (Peña, 1975; Ramos *et al.*, 1985). Consequently the $K^+$ concentration gradient across the cell membrane also represents a store of metabolic energy available for solute transport (Okorokov, 1985). $K^+$ efflux from cells frequently accompanies heavy metal uptake by fungi and yeasts. It can show a stoichiometry of approximately $1M^{2+}_{in}$ to $2K^+_{out}$ where M is a metal (Fuhrmann and Rothstein, 1968; Norris and Kelly, 1977; Lichko *et al.*, 1980; Mowll and Gadd, 1984; Gadd and Mowll, 1985), which is consistent with coupling between $K^+$ efflux and $M^{2+}$ uptake. However, $K^+$ efflux does not always accompany heavy metal uptake (Norris and Kelly, 1977; Mowll and Gadd, 1983; Okorokov *et al.*, 1983a) and, where it does occur, it may show no simple stoichiometry with metal uptake (Passow and Rothstein, 1960; Norris and Kelly, 1977; Gadd and Mowll, 1983; Kessels *et al.*, 1985; White and Gadd, 1987a,b). One major cause of non-stoichiometric $K^+$ efflux is toxicity, heavy metal ions causing extensive damage to cell membranes (Passow and Rothstein, 1960; Norris and Kelly, 1977; Mowll and Gadd, 1983). However, coupled $K^+$ efflux and $M^{2+}$ uptake can also occur in the absence of a $2K^+_{out}:1M^{2+}_{in}$ stoichiometry where $K^+$ efflux is also coupled to other energy-utilizing processes such as polyphosphate synthesis (Okorokov *et al.*, 1983a,b).

The retention of $K^+$ by fungi is in part an active process depending on the membrane potential and thus the $H^+$-translocating ATPase activity of the cell membrane (Ramos et al., 1985). Consequently, $K^+$ efflux can also result from treatment with ATPase inhibitors (Borst-Pauwels et al., 1983). Heavy metals can also act as inhibitors of the $H^+$-translocating ATPase repressing energy-dependent $H^+$ efflux and $K^+$ uptake by yeast cells (White and Gadd, 1987a,b). Heavy metal-induced $K^+$ efflux showed two phases, reversible at low metal ion concentrations and irreversible at high concentrations (White and Gadd, 1987a,b). Loss of $K^+$ by yeast may be an all-or-none effect in individual cells (Kuypers and Roomans, 1979; Theuvenet et al., 1983; Belde et al., 1988). Where populations of yeast cells were composed of sensitive cells, which lost $K^+$ in response to $Cd^{2+}$, and resistant cells, the latter subsequently took up the $K^+$ lost by the sensitive cells (Belde et al., 1988) indicating a mechanism for reversible $K^+$ loss.

It has been suggested that variations in the relationship of $K^+$ efflux to metal uptake represent different mechanisms of energization for ion transport (Okorokov, 1985) but it is clearly a more complex phenomenon, which also involves other processes that are the result of metabolic inhibition and structural damage by heavy metals.

*Kinetics of heavy metal transport.* The rate of heavy metal influx into yeast cells or mycelium is concentration-dependent and generally shows saturation with increasing external ion concentration. The kinetics frequently conform to the Michaelis–Menten model for enzyme-mediated reactions although the reported parameters vary greatly (*Table 1*) (Borst-Pauwels, 1981). One reason for such a departure is the existence of two or more transport systems of different affinities. Thus, uptake of $Co^{2+}$ by *S. cerevisiae* showed apparently biphasic kinetics (Norris and Kelly, 1977). However, reduction in surface potential due to cation binding at negatively charged sites may also modify the transport kinetics of alkali earth cations such as $Ca^{2+}$ and $Sr^{2+}$ (Roomans et al., 1979; Borst-Pauwels and Theuvenet, 1984) and heavy metal cations such as $Co^{2+}$ (Norris and Kelly, 1977). It has also been proposed that the reduction in surface potential could mimic the effects of saturation kinetics in the absence of a saturable transport system (Borst-Pauwels and Theuvenet, 1984).

Unlike the alkaline earth metals, even modest concentrations of many heavy metals can produce toxic disruptive effects on the structure and function of the cell membrane which may alter the kinetics of metal uptake. The inhibition of $H^+$ pumping by the membrane increased with metal concentration (White and Gadd, 1987b) and this could produce a progressive deenergization of the cell membrane which could affect the kinetics of a saturable transport system. At higher heavy metal concentrations the cell membrane permeability is increased leading to irreversible $K^+$ loss from the cell (Passow and Rothstein, 1960; Kuypers and Roomans, 1979; White and Gadd, 1987a,b). An apparent second phase of low-affinity transport coincided with this in one study (*Figure 2*) which suggested that, in this instance, biphasic kinetics was caused by cell membrane permeabilization (White and Gadd, 1987a). When different studies of heavy metal transport are compared, there is an apparent general trend that where the concentration range used is high the affinity obtained is low (*Table 1*). Where the concentrations (and often the apparent $K_m$ values) are within the range of potential toxicity, there is thus a clear possibility that these values may be artefacts.

Several attempts have been made to utilize formal studies of cation transport kinetics

**Table 1.** Some transport parameters for metal ion uptake by yeasts and fungi.

| Cation | Organism | $K_m$ (µM) | Conc. range (µM) | Reference |
|---|---|---|---|---|
| Co²⁺ | Saccharomyces cerevisiae | <10 | 10–250 | Fuhrmann and Rothstein (1968) |
| | Saccharomyces cerevisiae | 77 | <500 | Norris and Kelly (1977) |
| Ni²⁺ | Saccharomyces cerevisiae | 500 | <5000 | Fuhrmann and Rothstein (1968) |
| Cu²⁺ | Aureobasidium pullulans | 200 | 2–80 | Gadd and Mowll (1985) |
| | Penicillium ochro-chloron | 390 | 0–1000 | Gadd and White (1985) |
| Zn²⁺ | Saccharomyces cerevisiae | <10 | – | Fuhrmann and Rothstein (1968) |
| | Saccharomyces cerevisiae | 1300 | 500–7500 | Ponta and Broda (1970) |
| | Neocosmospora vasinfecta | 200 | – | Paton and Budd (1972) |
| | Neocosmospora vasinfecta | 1.1 | <10 | Budd (1988) |
| | | 770 | >10 | |
| | Candida utilis | 1.3–1.8[a] | 0.6–10.1 | Failla et al. (1976) |
| | Candida utilis | 2 | 1–10 | Failla and Weinberg (1977) |
| | Candida utilis | 0.36 | – | Lawford et al. (1980) |
| | Saccharomyces cerevisiae | 5000 | 10–500 | Mowll and Gadd (1983) |
| | Saccharomyces cerevisiae | 3.7 | <80 | White and Gadd (1987a) |
| | Aureobasidium pullulans (vegetative cells) | 10.8–11.8 | 2–20 | Gadd et al. (1987) |
| | (chlamydospore protoplasts) | 66.7 | | |
| Cd²⁺ | Saccharomyces cerevisiae | 1000 | – | Norris and Kelly (1977) |
| | Aureobasidium pullulans | 100–110 | 10–1000 | Mowll and Gadd (1984) |

[a]Non-Michaelis–Menten kinetics; the values given are concentrations giving half-maximal uptake.

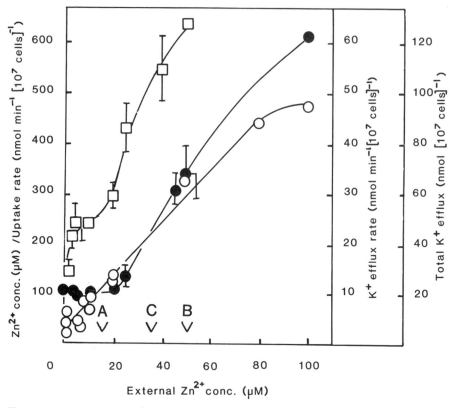

**Figure 2.** Zinc uptake by *Saccharomyces cerevisiae* (○), as a Woolf plot against concentration to show deviation from Michaelis−Menten kinetics. Total $K^+$ efflux (●), $K^+$ efflux rate (□) and the $Zn^{2+}$ concentrations yielding 50% (**A**) and 100% (**B**) inhibition of $K^+$ uptake and 50% inhibition of $H^+$ efflux (**C**) are shown for comparison. Data from White and Gadd (1987a).

to elucidate transport mechanisms. A central problem to this approach has been that many formally distinct models closely fit the observed relationships of transport rate to external ion concentration. For example $Ca^{2+}$ and $Sr^{2+}$ uptake by yeast could be described by means of two single-site carriers with respectively high and low affinity Michaelis−Menten kinetics or by means of either a single site carrier system with Michaelis−Menten kinetics or simple diffusion if the surface charge of the cell membrane was modified by cation binding (Roomans *et al.*, 1979; Borst-Pauwels and Theuvenet, 1984). Other theoretical studies have examined the possible role of the negative charges on the cell surface in modifying the kinetics of transport by increasing the effective cation concentration at the cell membrane. These have shown that saturation of these negative surface sites at increasing cation concentrations can result in kinetics of very similar form to the Michaelis−Menten model (Theuvenet and Borst-Pauwels, 1976; Borst-Pauwels, 1981). Uptake of $Rb^+$ was inhibited by $Ca^{2+}$, other cations and low pH with a direct relationship between the apparent $K_m$ and the surface charge as measured by the zeta potential (Gage *et al.*, 1985). Studies of the kinetics of carrier-mediated ion translocation have shown that there is a broad equivalence in the predicted rate/concentration relationship between several mechanisms, for instance between a dual carrier system and a single carrier with a regulatory binding site (Borst-Pauwels, 1973, 1974, 1976) and that further criteria are required to distinguish them (Borst-Pauwels, 1976). As a result of these conclusions, and also the further complications which may result from toxic effects, it seems unlikely that mechanistic questions concerning heavy metal transport can be answered on kinetic grounds alone.

*pH effects.* Accumulation of heavy metals by surface-binding increases with increasing pH (Gadd and Griffiths, 1980a; Gadd, 1986a). Energy-dependent heavy metal transport is also pH-dependent. In general, the rates of transport are maximal in the pH range 6.0−7.0 (Fuhrmann and Rothstein, 1968; Failla *et al.*, 1976; Roomans *et al.*, 1979; Mowll and Gadd, 1984; Gadd and Mowll, 1985). Copper uptake by *Penicillium ochro-chloron* differed in having low, constant uptake at between pH 3.0 and 5.5 but much greater uptake at pH 6.0 and above (Gadd and White, 1985).

The underlying mechanisms of these pH effects on heavy metal transport and binding are probably complex, reflecting variations in both the solution chemistry and speciation of heavy metals and the effects of pH on metal-binding sites at the cell surface. Most heavy metal cations can undergo both hydrolysis and polymerization in solution and both of these types of reaction are stimulated by raising the pH (Hunt, 1963). Such reactions were considered to be important in modifying thorium biosorption by *R.arrhizus* (Tzesos and Volesky, 1981) and copper uptake by *P.ochro-chloron* (Gadd and White, 1985). The negative surface potential is increased by dissociation of acidic groups at higher pH thus increasing the effective cation concentration at the membrane surface (Roomans *et al.*, 1979; Borst-Pauwels, 1981; Gage *et al.*, 1985). The internal pH of the cell can also affect monovalent and divalent cation transport with an optimum internal pH of about 6.5 (Theuvenet *et al.*, 1977; Roomans *et al.*, 1979; Borst-Pauwels, 1981) and is itself, to some extent, variable with the external pH (Borst-Pauwels, 1981).

*Heavy metal transport and resistance to toxicity.* Lowered uptake of heavy metals is a frequent mechanism of resistance. $Cd^{2+}$- and $Cu^{2+}$-resistant strains of *S.cerevisiae* showed reduced uptake of these metals (Joho *et al.*, 1983; Gadd *et al.*, 1984a). Similarly, a $Ni^{2+}$-resistant strain of *Neurospora crassa* showed reduced $Ni^{2+}$ uptake compared to sensitive strains (Mohan *et al.*, 1984). *P.ochro-chloron* is extremely tolerant of $Cu^{2+}$ and maintains a constant intracellular $Cu^{2+}$ content when the external $[Cu^{2+}]$ exceeds about 16 mM (Gadd *et al.*, 1984b). This organism was also resistant to $Zn^{2+}$ and showed similar low uptake of $Zn^{2+}$ at high concentrations. However, it was sensitive to both $Ni^{2+}$ and $Co^{2+}$ for which it possessed high uptake (Okamoto *et al.*, 1977). In the polymorphic fungus *A.pullulans*, chlamydospores show little or no intracellular uptake of heavy metals such as $Cu^{2+}$, $Cd^{2+}$, $Co^{2+}$ or $Zn^{2+}$ (Mowll and Gadd, 1984; Gadd and Mowll, 1985; Gadd *et al.*, 1987). This organism also produces chlamydospores in response to the presence of heavy metals (Gadd and Griffiths, 1980b) as well as under other unfavourable conditions and this response appears to be important in the success of the organism in metal-polluted habitats (Gadd, 1981).

However, reduced metal uptake is absent in some heavy metal-resistant strains of yeasts and fungi. For instance, some $Ni^{2+}$-resistant strains of *N.crassa* showed similar $Ni^{2+}$ uptake to sensitive strains (Mohan *et al.*, 1984). An $Mn^{2+}$-resistant strain of

29

*S.cerevisiae* also showed increased uptake of $Mn^{2+}$ compared to the sensitive parent strain (Bianchi *et al.*, 1981). Thus reduced uptake of heavy metals is a major mechanism of heavy metal resistance in fungi but is not universally present in resistant strains.

## Intracellular localization and binding of heavy metals

### Metallothioneins

A common metal-induced response in fungi is the intracellular synthesis of cysteine-rich metal-binding proteins called metallothioneins which have functions in metal detoxification and also in the storage and regulation of intracellular metal ion concentrations. The copper-induced metallothionein of *S.cerevisiae*, often referred to as Cu-MT or yeast MT ($M_r$ 6573), is inducible only by copper and not, for example, cadmium or zinc (Butt and Ecker, 1987). This metal-regulated genetic system is proving to be a powerful tool in pure and applied microbiology and the mechanisms involved are receiving intensive study. *S.cerevisiae* contains a single Cu-MT gene (*CUP1*) located on chromosome VIII (Butt and Ecker, 1987). Copper-resistant (CUP1$^r$) yeast strains contain 2 to 14 copies of the *CUP1* locus, which are tandemly repeated, and these strains can grow on media containing high [$Cu^{2+}$], e.g. 2 mM (Fogel *et al.*, 1983; White and Gadd, 1986; Butt and Ecker, 1987). Copper-sensitive (cup1$^s$) strains do not grow on media containing $>0.15$ mM $Cu^{2+}$ though inhibitory concentrations depend on the medium employed (Gadd, 1989b). Further culture of *S.cerevisiae* strains in the presence of increasing concentrations of $Cu^{2+}$ can lead to selection of hyper-resistant strains (White and Gadd, 1986) which appear disomic for chromosome VIII (Butt and Ecker, 1987). Copper-inducible metallothionein has also been documented in *N.crassa* (Beltramini and Lerch, 1983).

Inducible, structurally distinct cadmium-binding proteins are found in *Schizosaccharomyces pombe*. These 'phytochelatins' are composed of three amino acids, cysteine, glutamic acid and glycine, and are analogous to similar peptides found in plant cells after exposure to heavy metals. They can bind a variety of transition metals in addition to cadmium (Butt and Ecker, 1987).

Cadmium-binding proteins have also been detected in strains of *S.cerevisiae*. After exposure of a sensitive strain to cadmium, most internal cadmium was bound to insoluble cytosolic material whereas cadmium-binding proteins ($M_r$ $<30\,000$) were detected in the cytosol of resistant strains (Joho *et al.*, 1986).

### The role of the vacuole

The fungal vacuole serves as a store for substances that might interfere with metabolism such as amino acids (Wiemken and Nurse, 1973), polyphosphates (Okorokov *et al.*, 1980) and some enzymes, particularly hydrolases (Matile and Wiemken, 1976). Various cations such as $K^+$ and $Mg^{2+}$ (Okorokov *et al.*, 1980) and $Ca^{2+}$ (Ohsumi and Anraku, 1983; Eilam *et al.*, 1985) are also mainly located in the yeast vacuole. This localization maintains the cytosolic $Ca^{2+}$ concentration at a low level (Eilam *et al.*, 1985) and similar compartmentation of $Ca^{2+}$ occurs in *N.crassa* (Cornelius and Nakashima, 1987). Other metal cations are also compartmentalized in the vacuole such as $Mn^{2+}$ in *Saccharomyces carlsbergensis* (Okorokov *et al.*, 1977) and also $Zn^{2+}$ and $Co^{2+}$ in

*S. cerevisiae* (White and Gadd, 1987a,b). Intracellular $Cu^{2+}$ and $Cd^{2+}$ pools in *A. pullulans* showed biphasic efflux kinetics suggesting compartmentation (Mowll and Gadd, 1984; Gadd and Mowll, 1985). In *S. cerevisiae*, however, the main soluble fraction of both $Cd^{2+}$ and $Cu^{2+}$ was in the cytosol and the distribution was similar in sensitive and resistant strains (Joho *et al.*, 1985a,b; White and Gadd, 1986). In contrast, a $Co^{2+}$-resistant strain of yeast had a higher proportion of the intracellular $Co^{2+}$ in the vacuole than the sensitive parental strain (White and Gadd, 1986), which strongly suggests that the vacuole functioned to store it out of contact with metabolically active cell components. It appears that metals such as $Cd^{2+}$, which are not concentrated in the vacuole, are detoxified by other means such as sequestration by metal-binding proteins (see above). Chemical binding may also be combined with partitioning into the vacuole. For example, $Mn^{2+}$ was accumulated in the yeast vacuole with polyphosphate (Okorokov *et al.*, 1977; Lichko *et al.*, 1982) and vacuolar $Zn^{2+}$ was also associated with polyphosphate bodies in both yeast and filamentous fungi (Paton and Budd, 1972; Doonan *et al.*, 1979; Bilinski and Miller, 1983).

*Heavy metal transport by the vacuole.* The vacuolar membrane has an $H^+$ antiport system for basic amino acids and for $Ca^{2+}$ (Ohsumi and Anraku, 1983; Eilam *et al.*, 1985) and it appears that a similar system exists for those heavy metals that are preferentially located in the vacuole (Okorokov *et al.*, 1985). In the presence of ATP, yeast vacuolar preparations generated an inside-positive proton concentration gradient, which was dissipated by heavy metals such as $Zn^{2+}$ during uptake (Okorokov *et al.*, 1985). The efflux was greater than the $2H^+_{out}/1Ca^{2+}_{in}$ seen in $Ca^{2+}$ transport, apparently indicating a loose coupling of these fluxes. $Mn^{2+}$ was also transported into the vacuole by a similar system in *S. carlsbergensis* (Lichko *et al.*, 1982) with a $K_m$ for vacuolar transport very close to its observed cytosolic concentration (Okorokov *et al.*, 1985). $Zn^{2+}$ transport into isolated *S. cerevisiae* vacuoles was also consistent with an $H^+$ antiporter system (White and Gadd, 1987a).

As well as proton gradients, other ion gradients may be involved in energization of vacuolar transport as $Mn^{2+}$ taken up by *S. carlsbergensis* was partitioned into the vacuole with an efflux of $K^+$ from the vacuole (Lichko *et al.*, 1980).

## Regulation of heavy metal transport and localization

Many heavy metals are both physiologically essential and potentially toxic to fungi and yeasts. This determines many aspects of metal metabolism such as the existence of binding proteins or localization in the vacuole. These processes clearly require regulation and this has been demonstrated in some studies. *Candida utilis* grown in $Zn^{2+}$-limited chemostat culture showed greater $Zn^{2+}$ accumulation than carbon-limited cells (Lawford *et al.*, 1980) and $Zn^{2+}$ uptake was similarly enhanced in $Zn^{2+}$-limited *Neocosmospora vasinfecta* (Budd, 1988). Conversely, $Mn^{2+}$ uptake by *C. utilis* was reduced when it was grown in chemostat culture with an excess of $Mn^{2+}$ as compared with cells grown in unsupplemented medium. $Cu^{2+}$ uptake, however, was not regulated (Parkin and Ross, 1986). Limitation of uptake is a frequent response to high external heavy metal concentrations producing tolerance. In a strain of *S. cerevisiae* which was trained to $Co^{2+}$ tolerance, enhanced sequestration of $Co^{2+}$ in the vacuole appeared to be genetically determined while reduced $Co^{2+}$ transport appeared to be a physiological adaptation (White and Gadd, 1986).

The requirements of the organism for a particular ion may vary depending on its physiological state, for instance with the induction of different metalloenzymes during the cell cycle. Metal uptake and/or compartmentation may be regulated to match this. Batch cultures of *C. utilis* showed greater $Zn^{2+}$ uptake during the lag phase or the late exponential phase than in the exponential phase of growth (Failla and Weinberg, 1977) and in a strain of *Aspergillus parasiticus*, hyperaccumulation of $Zn^{2+}$ by depleted cultures was prevented by prior exposure to $Zn^{2+}$ during the period $20-30$ h post inoculum (Failla and Niehaus, 1986). The intracellular localization of metal ions may also be regulated. $Zn^{2+}$ was mainly located in the vacuolar polyphosphate bodies of *S. cerevisiae* in vegetative cells while it was concentrated in the nucleus during ascosporogenesis (Bilinski and Miller, 1983).

## Biotechnological aspects of metal—fungal interactions

There are applied aspects to several areas of metal—fungal interactions. Metal toxicity is the basis of many antifungal preparations for the control of animal and plant pathogens and the preservation of natural and man-made materials. Further knowledge of mechanisms and assessment of toxicity is essential for the formulation and application of more effective compounds (Gadd, 1986c). The use of mycorrhizas to assist land reclamation may also be a commercial possibility (Bradley *et al.*, 1982).

The removal of heavy metals and radionuclides from solution by fungal biomass is a topic of current interest and this can lead to detoxification and/or the recovery of valuable metals, e.g. gold from waste waters and effluents (Gadd, 1986b, 1989a,b,c; Tsezos, 1986; Gadd *et al.*, 1988). As described, both living and dead fungal biomass as well as derived or excreted products can remove heavy metals and radionuclides from solution and there is no doubt that certain biomass types are highly efficient with capacities frequently greater than commercially-obtained ion-exchange resins (Gadd, 1989b,c). Microbe-based technologies may provide an alternative or supplement to conventional treatment methods and a variety of systems are already in commercial use in the mining and metallurgical industries (Hutchins *et al.*, 1986; Gadd, 1989c). However, many processes are relatively unsophisticated and many areas of metal—fungal interactions remain unexploited.

Toxicity may inhibit industrial applications of living cell systems although it may be possible to separate cell propagation from the metal contacting phase or to use resistant strains. Clearly, resistance is commonly found in fungi and yeasts. The use of dead biomass or derived products eliminates problems of toxicity, nutrient supply and maintenance of optimal growth and metal recovery and biomass regeneration may be relatively easy by means of non-destructive desorption using, e.g. carbonates or dilute acids. However, living cells exhibit a wider variety of mechanisms for metal accumulation, for example transport, intracellular and extracellular precipitation etc. although, as described, decreased uptake may often be associated with resistance and internalized metals may be difficult to recover unless destructive treatments, for example, incineration, are used.

For rigorous industrial applications, freely suspended microbial biomass has a number of disadvantages such as small particle size and low density and mechanical strength which can make biomass/effluent separation difficult. For use in packed-bed or fluidized-bed bioreactors, immobilized or pelleted biomass, which can be living or dead, may have more potential. Immobilized biomass, whether within or on an inert matrix, has the advantages of easy separation of cells and effluent, high flow rates, with and/or without recirculation, minimal clogging, high biomass loadings and better capabilities for manipulation and re-use (Gadd, 1986b; Tsezos, 1986). Uranium removal has been demonstrated using biomass particles immobilized using various polymeric membranes and such preparations may be capable of >99% metal removal from dilute solutions (Tsezos, 1986; Hutchins *et al.*, 1986). Other examples include *Aspergillus oryzae* immobilized on reticulated foam (Kiff and Little, 1986) and *Trichoderma viride* packed in molochite and used for copper removal (Townsley *et al.*, 1986). Several filamentous fungi can be grown in pellet form and these may have similar advantages to immobilized preparations. Pellets of *Aspergillus niger* were capable of efficient uranium removal when used in a fluidized-bed reactor. However, the similarity in density between cells and the liquid medium made continuous operation difficult (Yakubu and Dudeney, 1986). It is generally accepted that for improved industrial use, immobilized or pelleted preparations should be used with recovery utilizing a cheap stripping agent which can be recycled (Brierley *et al.*, 1986; Tsezos, 1986). Where recovery is not necessarily desired, for example of non-precious metals or some radionuclides, processes should be devised that produce low-volume containable waste (Gadd, 1986b).

Yeast metallothionein may have application in metal recovery since it can bind other metals besides copper, e.g. cadmium, zinc, cobalt, silver and gold, although these metals cannot induce MT synthesis. It may be possible to develop yeast strains with constitutive expression of MT genes which may be high accumulators or to engineer systems for the secretion of MT proteins with specific affinities for different metals (Butt and Ecker, 1987).

It is clear that some examples of metal recovery are competitive with conventional treatments (Brierley *et al.*, 1986) but further work is needed in several areas, e.g. transport, particulate metal accumulation, extracellular precipitation, complexation and metallothioneins to realize the full potential of fungal systems. It is desirable that administrative authorities and those industries responsible for the deliberate and accidental release of metals or radionuclides into the environment take a prominent lead in the support and development of such work.

# References

Adjimani,J.P. and Emery,T. (1987) Iron uptake in *Mycelia sterilia* EP-76. *Journal of Bacteriology* **169**, 3664–3668.

Belde,P.J.M., Kessels,B.G.F., Moelans,I.M. and Borst-Pauwels,G.W.F.H. (1988) $Cd^{2+}$ uptake, $Cd^{2+}$ binding and loss of cell $K^+$ by a Cd-sensitive and a Cd-resistant strain of *Saccharomyces cerevisiae*. *FEMS Microbiology Letters* **49**, 493–498.

Beltramini,M. and Lerch,K. (1983) Spectroscopic studies on *Neurospora* copper metallothionein. *Biochemistry* **22**, 2043–2048.

Bianchi,M.E., Carbone,M.L. and Lucchini,G. (1981) $Mn^{2+}$ and $Mg^{2+}$ uptake in Mn-sensitive and Mn-resistant yeast strains. *Plant Science Letters* **22**, 345–352.

Bilinski,C.A. and Miller,J.J. (1983) Translocation of zinc from vacuole to nucleus during yeast meiosis. *Canadian Journal of Cytology and Genetics* **25**, 415−419.

Blunden,S.J., Hobbs,L.A. and Smith,P.J. (1984) The environmental chemistry of organotin compounds. In *Environmental Biochemistry* (ed. H.J.M.Bowen) Royal Society of Chemistry, London.

Borst-Pauwels,G.W.F.H. (1973) Two site − single carrier transport kinetics. *Journal of Theoretical Biology* **40**, 19−31.

Borst-Pauwels,G.W.F.H. (1974) Multi-site two-carrier transport comparison with enzyme kinetics. *Journal of Theoretical Biology* **48**, 183−195.

Borst-Pauwels,G.W.F.H. (1976) Discrimination criteria for apparent two-site transport models. *Journal of Theoretical Biology* **56**, 191−204.

Borst-Pauwels,G.W.F.H. (1981) Ion transport in yeast. *Biochimica et Biophysica Acta* **650**, 88−127.

Borst-Pauwels,G.W.F.H. and Theuvenet,A.P.R. (1984) Apparent saturation kinetics of divalent cation uptake in yeast caused by a reduction in the surface potential. *Biochimica et Biophysica Acta* **771**, 171−176.

Borst-Pauwels,G.W.F.H., Theuvenet,A.P.R. and Stols,A.L.H. (1983) All-or-none reactions of inhibitors of the plasma membrane ATPase with *Saccharomyces cerevisiae*. *Biochimica et Biophysica* **732**, 186−192.

Bowman,B.J., Mainzer,S.E., Allen,K.E. and Slayman,C.W. (1978) Effects of inhibitors on the plasma membrane and mitochondrial adenosine-triphosphatases of *Neurospora crassa*. *Biochimica et Biophysica Acta* **512**, 13−28.

Bradley,R., Burt,A.J. and Read,D.J. (1982) The biology of mycorrhiza in the Ericaceae. VIII. The role of mycorrhizal infection in heavy metal resistance. *New Phytologist* **91**, 197−209.

Brierley,J.A., Goyak,G.M. and Brierley,C.L. (1986) Considerations for commercial use of natural products for metals recovery. In *Immobilisation of Ions by Bio-sorption* (eds H.Eccles and S.Hunt), Ellis Horwood, Chichester, pp. 105−117.

Budd,K. (1988) A high-affinity system for the transport of zinc in *Neocosmospora vasinfecta*. *Experimental Mycology* **12**, 195−202.

Butt,T.R. and Ecker,D.J. (1987) Yeast metallothionein and applications in biotechnology. *Microbiological Reviews* **51**, 351−364.

Cooney,J.J. and Wuertz,S. (1989) Toxic effects of tin compounds on microorganisms. *Journal of Industrial Microbiology* (in press).

Cooney,J.J., de Rome,L., Laurence,O.S. and Gadd,G.M. (1989) Effects of organotins and organoleads on yeasts. *Journal of Industrial Microbiology* (in press).

Cooper,T.G. (1982) Transport in *Saccharomyces cerevisiae*. In *The Molecular Biology of the Yeast Saccharomyces* Vol. 2. *Metabolism and Gene Expression* (eds J.N.Strathern, E.W.Jones and J.R.Broach), Cold Spring Harbor Laboratory, New York, pp. 399−461.

Cornelius,G. and Nakashima,H. (1987) Vacuoles play a decisive role in calcium homeostasis in *Neurospora crassa*. *Journal of General Microbiology* **133**, 2341−2347.

Doonan,B.B., Crang,R.E., Jensen,T.E. and Baxter,M. (1979) *In situ* X-ray dispersive microanalysis of polyphosphate bodies in *Aureobasidium pullulans*. *Journal of Ultrastructure Research* **69**, 232−238.

Duddridge,J.E. and Wainwright,M. (1980) Heavy metal accumulation by aquatic fungi and reduction in viability of *Gammarus pulex* fed $Cd^{2+}$ contaminated mycelium. *Water Research* **14**, 1605−1611.

De Rome,L. and Gadd,G.M. (1987) Copper adsorption by *Rhizopus arrhizus, Cladosporium resinae* and *Penicillium italicum*. *Applied Microbiology and Biotechnology* **26**, 84−90.

Duxbury,T. (1985) Ecological aspects of heavy metal responses in microorganisms. In: *Advances in Microbial Ecology* (Ed. K.C.Marshall), Plenum Press, New York, pp. 185−235.

Eilam,Y., Lavi,H. and Grossowicz,N. (1985) Cytoplasmic $Ca^{2+}$ homeostasis maintained by a vacuolar $Ca^{2+}$ transport system in the yeast *Saccharomyces cerevisiae*. *Journal of General Microbiology* **131**, 623−629.

Failla,L.J. and Niehaus,W.G. (1986) Regulation of $Zn^{2+}$ uptake and versicolorin A synthesis in a mutant strain of *Aspergillus parasiticus*. *Experimental Mycology* **10**, 35−41.

Failla,M.L. and Weinberg,E.D. (1977) Cyclic accumulation of zinc by *Candida utilis* during growth in batch culture. *Journal of General Microbiology* **99**, 85−97.

Failla,M.L., Benedict,C.D. and Weinberg,D. (1976) Accumulation and storage of $Zn^{2+}$ by *Candida utilis*. *Journal of General Microbiology* **94**, 23−36.

Fogel,S., Welch,J.W. and Karin,M. (1983) Gene amplification in yeast; CUP1 copy number regulates copper resistance. *Current Genetics* **7**, 1−9.

Fuhrmann,G.-F. and Rothstein,A. (1968) The transport of $Zn^{2+}$, $Co^{2+}$ and $Ni^{2+}$ into yeast cells. *Biochimica et Biophysica Acta* **463**, 325−330.

Gadd,G.M. (1981) Mechanisms implicated in the ecological success of polymorphic fungi in metal-polluted habitats. *Environmental Technology Letters* **2**, 531−536.

Gadd,G.M. (1986a) Fungal responses towards heavy metals. In *Microbes in Extreme Environments* (eds G.A.Codd and R.A.Herbert), Academic Press, London, pp. 83–110.

Gadd,G.M. (1986b) The uptake of heavy metals by fungi and yeasts: the chemistry and physiology of the process and applications for biotechnology. In *Immobilisation of Ions by Bio-sorption* (eds H.Eccles and S.Hunt), Ellis Horwood, Chichester, pp. 135–147.

Gadd,G.M. (1986c) Toxicity screening using fungi and yeasts. In *Toxicity Testing using Microorganisms* (eds B.J.Dutka and G.Bitton), Vol. 2, CRC Press, Boca Raton, FL, pp. 43–77.

Gadd,G.M. (1989a) Metal tolerance. In *Extremophiles* (ed. C.Edwards), Open University Press, Milton Keynes (in press).

Gadd,G.M. (1989b) Fungi and yeasts for metal accumulation. In *Microbial Mineral Recovery* (eds H.L.Ehrlich, J.A.Brierley and C.L.Brierley), MacMillan, New York (in press).

Gadd,G.M. (1989c) Accumulation of metals by microorganisms and algae. In *Biotechnology—A Comprehensive Treatise* (eds H.-J.Rehm and G.Reed), Vol. 6b, VCH Verlagsgesellschaft, Weinheim, pp. 401–433.

Gadd,G.M. and de Rome,L. (1988) Biosorption of copper by fungal melanin. *Applied Microbiology and Biotechnology* **29**, 610–617.

Gadd,G.M. and Edwards,S.W. (1986) Heavy metal-induced flavin production by *Debaryomyces hansenii* and possible connexions with iron metabolism. *Transactions of the British Mycological Society* **87**, 533–542.

Gadd,G.M. and Griffiths,A.J. (1980a) Influence of pH on toxicity and uptake of copper in *Aureobasidium pullulans*. *Transactions of the British Mycological Society* **75**, 91–96.

Gadd,G.M. and Griffiths,A.J. (1980b) Effect of copper on morphology of *Aureobasidium pullulans*. *Transactions of the British Mycological Society* **74**, 387–392.

Gadd,G.M. and Mowll,J.L. (1983) The relationship between cadmium uptake, potassium release and viability in *Saccharomyces cerevisiae*. *FEMS Microbiology Letters* **16**, 45–48.

Gadd,G.M. and Mowll,J.L. (1985) Copper uptake by yeast-like cells, hyphae and chlamydospores of *Aureobasidium pullulans*. *Experimental Mycology* **9**, 230–240.

Gadd,G.M. and White,C. (1985) Copper uptake by *Penicillium ochro-chloron*: influence of pH on toxicity and demonstration of energy-dependent copper influx using protoplasts. *Journal of General Microbiology* **131**, 1875–1879.

Gadd,G.M., Stewart,A., White,C. and Mowll,J.L. (1984a) Copper uptake by whole cells and protoplasts of a wild-type and copper-resistant strain of *Saccharomyces cerevisiae*. *FEMS Microbiology Letters* **24**, 231–234.

Gadd,G.M., Chudek,J.A., Foster,R. and Reed,R.H. (1984b) The osmotic responses of *Penicillium ochro-chloron*: changes in internal solute levels in response to copper and salt stress. *Journal of General Microbiology* **130**, 1969–1975.

Gadd,G.M., White,C. and Mowll,J.L. (1987) Heavy metal uptake by intact cells and protoplasts of *Aureobasidium pullulans*. *FEMS Microbiology Ecology* **45**, 261–267.

Gadd,G.M., White,C. and de Rome,L. (1988) Heavy metal and radionuclide uptake by fungi and yeasts. In *Biohydrometallurgy* (eds P.R.Norris and D.P.Kelly), Science and Technology Letters, Kew, pp. 421–435.

Gage,R.A., van Wijngarden,W., Theuvenet,A.P.R., Borst-Pauwels,G.W.F.H. and Verkleij,A.J. (1985) Inhibition of Rb$^+$ uptake in yeast by Ca$^{2+}$ is caused by reduction in the surface potential and not the Donnan potential of the cell wall. *Biochimica et Biophysica Acta* **812**, 1–8.

Galun,M., Keller,P., Feldstein,H., Galun,E., Siegel,S. and Siegel,B. (1983a) Recovery of uranium (IV) from solution using fungi II. Release from uranium-loaded *Penicillium* biomass. *Water, Air and Soil Pollution* **20**, 277–285.

Galun,M., Keller,P., Malki,D., Feldstein,H., Galun,E., Siegel,S. and Siegel,B. (1983b) Removal of uranium (VI) from solution by fungal biomass: inhibition by iron. *Water, Air and Soil Pollution* **21**, 411–414.

Goffeau,A. and Slayman,C.W. (1981) The proton-translocating ATPase of the fungal plasma membrane. *Biochimica et Biophysica Acta* **639**, 197–223.

Greenaway,W. (1972) Permeability of phenyl-Hg$^+$-resistant and phenyl-Hg$^+$-susceptible isolates of *Pyrenophora avenae* to the phenyl-Hg$^+$. *Journal of General Microbiology* **73**, 251–255.

Hockertz,S., Schmidt,J. and Auling,G. (1987) A specific transport system for manganese in the filamentous fungus *Aspergillus niger*. *Journal of General Microbiology* **133**, 3513–3519.

Hunt,J.P. (1963) *Metal Ions in Aqueous Solution*. W.A.Benjamin, New York.

Hutchins,S.R., Davidson,M.S., Brierley,J.A. and Brierley,C.L. (1986) Microorganisms in reclamation of metals. *Annual Review of Microbiology* **40**, 311–336.

Joho,M., Sukendbu,Y., Egashira,E. and Murayama,T. (1983) The correlation between Cd$^{2+}$ sensitivity and Cd$^{2+}$ uptake in the strains of *Saccharomyces cerevisiae*. *Plant and Cell Physiology* **24**, 389–394.

Joho,M., Imai,M. and Murayama,T. (1985a) Different distribution of $Cd^{2+}$ between Cd-sensitive and Cd-resistant strains of *Saccharomyces cerevisiae*. *Journal of General Microbiology* **131**, 53−56.

Joho,M., Fujioka,Y. and Murayama,T. (1985b) Further studies on the subcellular distribution of $Cd^{2+}$ in Cd-sensitive and Cd-resistant strains of *Saccharomyces cerevisiae*. *Journal of General Microbiology* **131**, 3185−3191.

Joho,M., Yamanaka,C. and Murayama,T. (1986) $Cd^{2+}$ accommodation by *Saccharomyces cerevisaie*. *Microbios* **45**, 169−179.

Kiff,R.J. and Little,D.R. (1986) Biosorption of heavy metals by immobilized fungal biomass. In *Immunobilisation of Ions by Bio-sorption* (eds H.Eccles and S.Hunt), Ellis Horwood, Chichester, pp. 71−80.

Kessels,B.G.F., Belde,P.J.M. and Borst-Pauwels,G.W.F.H. (1985) Protection of *Saccharomyces cerevisiae* against $Cd^{2+}$ toxicity by $Ca^{2+}$. *Journal of General Microbiology* **131**, 2533−2537.

Kuypers,G.A.J.O. and Roomans,G.M. (1979) Mercury-induced loss of $K^+$ from yeast cells investigated by electron probe X-ray microanalysis. *Journal of General Microbiology* **115**, 13−18.

Lawford,H.G., Pik,J.R., Lawford,G.R., Williams,T. and Kligerman,A. (1980) Hyperaccumulation of zinc by zinc-depleted *Candida utilis* grown in chemostat culture. *Canadian Journal of Microbiology* **26**, 71−76.

Lichko,L.P., Okorokov,L.A. and Kulaev,I.S. (1980) Role of vacuolar ion pool in *Saccharomyces carlsbergensis*. Potassium efflux from vacuoles is coupled with manganese or magnesium influx. *Journal of Bacteriology* **144**, 666−671.

Lichko,L.P., Okorokov,L.A. and Kulaev,I.S. (1982) Participation of vacuoles in regulation of levels of $K^+$, $Mg^{2+}$ and orthophosphate ions in cytoplasm of the yeast *Saccharomyces carlsbergensis*. *Archives of Microbiology* **132**, 289−293.

Matile,P.H. and Wiemken,A. (1976) Interactions between cytoplasm and vacuole. In *Transport in Plants*. Vol. III. *Intracellular Interactions and Transport Processes* (eds C.R.Stocking and U.Heser), Springer-Verlag, Berlin, pp. 255−287.

McCord,J.M. and Day,D. (1978) Superoxide-dependent production of hydroxyl radical catalysed by iron-EDTA complex. *FEBS Letters* **86**, 139−142.

Mehlhorn,R.J. (1986) The interaction of inorganic species with biomembranes In *The Importance of Chemical Speciation in Environmental Processes* (eds M.Bernard, F.E.Brinckman and P.J.Sadler), Springer-Verlag, Berlin, pp. 85−97.

Minney,S.F. and Quirk,A.V. (1985) Growth and adaptation of *Saccharomyces cerevisiae* at different cadmium concentrations. *Microbios* **42**, 37−44.

Mohan,P.M., Rudra,M.P.P. and Sastry,K.S. (1984) Nickel transport in nickel-resistant strains of *Neurospora crassa*. *Current Microbiology* **10**, 125−128.

Mowll,J.L. and Gadd,G.M. (1983) Zinc uptake and toxicity in the yeasts *Sporobolomyces roseus* and *Saccharomyces cerevisiae*. *Journal of General Microbiology* **129**, 3421−3425.

Mowll,J.L. and Gadd,G.M .(1984) Cadmium uptake by *Aureobasidium pullulans*. *Journal of General Microbiology* **130**, 279−284.

Murphy,R.J. and Levy,J.F. (1983) Production of copper oxalate by some copper tolerant fungi. *Transactions of the British Mycological Society* **81**, 165−168.

Muzzarelli,R.A.A., Bregani,F. and Sigon,F. (1986) Chelating abilities of amino acid glucans and sugar acid glucans derived from chitosan. In *Immobilisation of Ions by Bio-sorption* (eds H.Eccles and S.Hunt), Ellis Horwood, Chichester, pp. 173−182.

Norris,P.R. and Kelly,D.P. (1977) Accumulation of cadmium and cobalt by *Saccharomyces cerevisiae*. *Journal of General Microbiology* **99**, 317−324.

Ochiai,E.I. (1987) *General Principles of Biochemistry of the Elements*. Plenum Press, New York.

Ohsumi,Y. and Anraku,Y. (1983) Calcium transport driven by proton motive force in vacuolar membrane vesicles of *Saccharomyces cerevisiae*. *Journal of Biological Chemistry* **258**, 5614−5617.

Okamoto,K., Suzuki,M., Toda,S. and Fuwa,K. (1977) Uptake of heavy metals by a copper-tolerant fungus, *Penicillium ochro-chloron*. *Agricultural and Biological Chemistry* **41**, 17−22.

Okorokov,L.A. (1985) Main mechanisms of ion transport and regulation of ion concentrations in the yeast cytoplasm. In *Environmental Regulation of Microbial Metabolism* (eds I.S.Kulaev, E.A.Dawes and D.W.Tempest), Academic Press, London, pp. 339−349.

Okorokov,L.A., Lichko,L.P., Kodomtseva,V.M., Kholodenko,V.P., Titovsky,V.T. and Kulaev,I.S. (1977) Energy-dependent transport of manganese into yeast cells and distribution of accumulated ions. *European Journal of Biochemistry* **75**, 373−377.

Okorokov,L.A., Lichko,L.P. and Kulaev,I.S. (1980) Vacuoles: main compartments of potassium, magnesium and phosphate ions in *Saccharomyces carlsbergensis* cells. *Journal of Bacteriology* **144**, 661−665.

Okorokov,L.A., Andreeva,N.A., Lichko,L.P. and Valiakhmetov,A.Y. (1983a) Transmembrane gradient of $K^+$ ions as an energy source in the yeast *Saccharomyces carlsbergensis*. *Biochemistry International* **6**, 463−472.

Okorokov,L.A., Lichko,L.P. and Andreeva,N.A. (1983b) Changes of ATP, polyphosphate and $K^+$ content in *Saccharomyces cerevisiae* during uptake of $Mn^{2+}$ and glucose. *Biochemistry International* **6**, 481−488.

Okorokov,L.A., Kulakovskaya,T.V., Lichko,L.P. and Polorotova,E.V. (1985) $H^+$/ion antiport as principal mechanism of transport systems in the vacuolar membrane of the yeast *Saccharomyces carlsbergensis*. *FEBS Letters* **192**, 303−306.

Oscik,J. (1982) *Adsorption*. Ellis Horwood, Chichester.

Parkin,M.J. and Ross,I.S. (1986) The regulation of $Mn^{2+}$ and $Cu^{2+}$ uptake by cells of the yeast *Candida utilis* grown in continuous culture. *FEMS Microbiology Letters* **37**, 59−62.

Passow,H. and Rothstein,A. (1960) The binding of mercury by yeast cells in relation to changes in permeability. *Journal of General Physiology* **43**, 621−633.

Paton,W.H.N. and Budd,K. (1972) Zinc uptake in *Neocosmospora vasinfecta*. *Journal of General Microbiology* **72**, 173−184.

Peña,A. (1975) Studies on the mechanism of $K^+$ transport in yeast. *Archives of Biochemistry and Biophysics* **167**, 397−409.

Ponta,H. and Broda,E. (1970) Mechanismen der aufnahme von zink durch backerhefe. *Planta* **95**, 18−26.

Ramos,S., Peña,P., Valle,E., Bergillos,L., Parra,F. and Lazo,P.S. (1985) Coupling of protons and potassium gradients in yeast. In *Environmental Regulation of Microbial Metabolism* (eds I.S.Kulaev, E.A.Dawes and D.W.Tempest), Academic Press, London, pp. 351−357.

Roomans,G.M., Theuvenet,A.P.R., van den Berg,Th.P.R. and Borst-Pauwels,G.W.F.H. (1979) Kinetics of $Ca^{2+}$ and $Sr^{2+}$ uptake by yeast. Effects of pH, cations and phosphate. *Biochimica et Biophysica Acta* **551**, 187−196.

Ross,I.S. and Walsh,A.L. (1981) Resistance to copper in *Saccharomyces cerevisiae*. *Transactions of the British Mycological Society* **77**, 27−32.

Ross,I.S. and Townsley,C.C. (1986) The uptake of heavy metals by filamentous fungi. In *Immobilisation of Ions by Bio-sorption* (eds H.Eccles and S.Hunt), Ellis Horwood, Chichester, pp. 49−58.

Saiz-Jimenez,C. and Shafizadeh,F. (1984) Iron and copper binding by fungal phenolic polymers: an electron spin resonance study. *Current Microbiology* **10**, 281−286.

Sakaguchi,T. and Nakajima,A. (1982) Recovery of uranium by chitin phosphate and chitosan phosphate. In *Chitin and Chitosan* (eds S.Mirano and S.Tokura), Japanese Society of Chitin and Chitosan, Tottori Japan, pp. 177−182.

Serrano,R. (1985) *Plasma Membrane ATPases of Plants and Fungi*. CRC Press, Boca Raton, FL.

Shumate,S.E. and Strandberg,G.W. (1985) Accumulation of metals by microbial cells. In *Comprehensive Biotechnology* (eds M.Moo-Young, C.N.Robinson and J.A.Howell), Vol. 4, Pergamon Press, New York, pp. 235−247.

Slayman,C.L. (1970) Movement of ions and electrogenesis in microorganisms. *American Zoologist* **10**, 377−392.

Slayman,C.L., Long,W.S. and Lu,C.Y.-H. (1973) The relationship between ATP and an electrogenic pump in the plasma membrane of *Neurospora crassa*. *Journal of Membrane Biology* **14**, 305−338.

Strandberg,G.W., Shumate,S.E. and Parrott,J.R. (1981) Microbial cells as biosorbents for heavy metals: accumulation of uranium by *Saccharomyces cerevisiae* and *Pseudomonas aeruginosa*. *Applied and Environmental Microbiology* **41**, 237−245.

Theuvenet,A.P.R. and Borst-Pauwels,G.W.F.H. (1976) The influence of surface charge on the kinetics of ion-translocation across biological membranes. *Journal of Theoretical Biology* **57**, 313−329.

Theuvenet,A.P.R., Roomans,G.M. and Borst-Pauwels,G.W.F.H. (1977) Intracellular pH and the kinetics of $Rb^+$ uptake by yeast, non-carrier versus carrier mediated uptake. *Biochimica et Biophysica Acta* **469**, 272−280.

Theuvenet,A.P.R. and Bindels,R.J.M. (1980) An investigation into the feasibility of using yeast protoplasts to study the ion transport properties of the plasma membrane. *Biochimica et Biophysica Acta* **599**, 587−595.

Theuvenet,A.P.R., Bindels,R.J.M., van Amelsvoort,J.M.N., Borst-Pauwels,G.W.F.H. and Stols,A.L.H. (1983) Interaction of ethidium bromide with yeast cells investigated by electron probe X-ray microanalysis. *Journal of Membrane Biology* **73**, 131−136.

Tobin,J.M., Cooper,D.G. and Neufeld,R.J. (1984) Uptake of metal ions by *Rhizopus arrhizus* biomass. *Applied and Environmental Microbiology* **47**, 821−824.

Townsley,C.C., Ross,I.S. and Atkins,A.S. (1986) Copper removal from a simulated leach effluent using the filamentous fungi *Trichoderma viride*. In *Immobilisation of Ions by Bio-sorption* (eds H.Eccles and S.Hunt), Ellis Horwood, Chicester, pp. 159−170.

Townsley,C.C. and Ross,I.S. (1986) Copper uptake in *Aspergillus niger* during batch growth and in non-growing mycelial suspensions. *Experimental Mycology* **10**, 281−288.

Tsezos,M. (1983) The role of chitin in uranium adsorption by *R.arrhizus*. *Biotechnology and Bioengineering* **25**, 2025−2040.

Tsezos,M. (1986) Adsorption by microbial biomass as a process for removal of ions from process or waste solutions. In *Immobilisation of Ions by Bio-sorption* (eds H.Eccles and S.Hunt), Ellis Horwood, Chichester, pp. 201−218.

Tsezos,M. and Volesky,B. (1981) Biosorption of uranium and thorium. *Biotechnology and Bioengineering* **22**, 583−604.

Tsezos,M. and Volesky,B. (1982a) The mechanism of uranium biosorption by *Rhizopus arrhizus*. *Biotechnology and Bioengineering* **24**, 385−401.

Tsezos,M. and Volesky,B. (1982b) The mechanism of thorium biosorption by *Rhizopus arrhizus*. *Biotechnology and Bioengineering* **24**, 955−969.

White,C. and Gadd,G.M. (1986) Uptake and cellular distribution of copper, cobalt and cadmium in strains of *Saccharomyces cerevisiae* cultured on elevated concentrations of these metals. *FEMS Microbiology Ecology* **38**, 277−283.

White,C. and Gadd,G.M. (1987a) The uptake and cellular distribution of zinc in *Saccharomyces cerevisiae*. *Journal of General Microbiology* **133**, 727−737.

White,C. and Gadd,G.M. (1987b) Inhibition of $H^+$ efflux and $K^+$ uptake and induction of $K^+$ efflux in yeast by heavy metals. *Toxicity Assessment* **2**, 437−447.

White,C. and Gadd,G.M. (1989) The removal of thorium from simulated acid process streams by fungal biomass. *Biotechnology and Bioengineering* (in press).

Wiemken,A. and Nurse,P. (1973) Isolation and characterisation of the amino-acid pools located within the cytoplasm and vacuoles of *Candida utilis*. *Planta* **109**, 293−306.

Wood,J.M. and Wang,H.K. (1983) Microbial resistance to heavy metals. *Environmental Science and Technology* **17**, 582−590.

Yakubu,N.A. and Dudeney,A.W.L. (1986) Biosorption of uranium with *Aspergillus niger*. In: *Immobilisation of Ions by Bio-sorption* (eds H.Eccles and S.Hunt), Ellis Horwood, Chichester, pp. 183−200.

Yakubu,N.A. and Dudeney,A.W.L. (1986) Biosorption of uranium with *Aspergillus niger*. In *Immobilisation of Ions by Bio-sorption* (eds H.Eccles and S.Hunt), Ellis Horwood, Chichester, pp. 183−200.

# CHAPTER 3

# Cadmium accumulation and resistance mechanisms in bacteria

MICHAEL H.RAYNER and PETER J.SADLER

*Department of Chemistry, Birkbeck College, University of London, Gordon House, 29 Gordon Square, London WC1H 0PP, UK*

## Cadmium in the environment

Cadmium is prevalent in the environment, being present in the Earth's crust at concentrations between 0.1 and 0.5 p.p.m., mostly as the sulphide. Cadmium finds widespread industrial use in electroplating, batteries, alloys, pigments, stabilizers for catalysts, and in semiconductors and TV tube phosphors. Since most zinc ores also contain cadmium it has a large turnover as a by-product from zinc refining.

Local cadmium concentrations can be very high as a result of natural deposits or industrial processes. Sewage sludge, for example, can contain 75 p.p.m. cadmium and its use on agricultural land has given typical soil concentrations of cadmium in the range 1.5–14.1 p.p.m. ($\sim 15 - 150 \mu M$) in areas of semi-rural London (Sherlock, 1983).

Our studies began with the view that the route of entry of cadmium into the food chain would depend on the chemical species in which it was found. It seemed likely that microorganisms play a major role in determining speciation and it was therefore important to further our understanding of how cadmium is metabolized by bacteria. In particular, we wished to investigate whether sulphur-rich metallothionein (MT)-type proteins play a role in bacteria as they do in mammals and other organisms (Kägi and Kojima, 1987). It might be assumed that cadmium exerts only toxic effects towards bacteria, but this cannot be assumed without appropriate evidence.

*Toxicity and essentiality*

Little work appears to have been done on the establishment of the essentiality of certain elements for bacterial growth; this is surprising. We do not know of any reported experiments which demonstrate that trace amounts of cadmium are not essential for bacterial growth. Such investigations have been carried out with warm-blooded mammals (Birch and Sadler, 1979), but they are expensive since all components of the diet must be carefully purified and the living environment defined. As a result of this some 25 elements are now thought to be essential for animal life, and Schwarz (1977) has classified three others, cadmium, arsenic and lead, as 'potentially essential'. Typically, 0.2 p.p.m. supplements of cadmium sulphate, a cadmium level close to that found normally in foods, caused a 13% growth increase for rats, and improved their appearance (Schwarz and Spallholz, 1978).

It is not just the element that matters, it is the level of administration (all compounds will be toxic at high concentrations) and the chemical form (Salder *et al.*, 1985; Bernhard *et al.*, 1986). For example, cadmium MT is a much more potent nephrotoxin than are aqua cadmium complexes (Chen and Ganther, 1975). The other components of the diet and the biochemical status of the animal are also important. Demonstration of essentiality requires the definition of specific biochemical roles in which the element is involved. For cadmium, none are known yet although cadmium-containing proteins and enzymes have been studied.

*Proteins and enzymes*

The best characterized cadmium proteins are the MTs, first isolated from equine kidney cortex and now known throughout the animal kingdom. They have a low $M_r$ (6−7 k) and high sulphur content (11%) from cysteines, which comprise about one-third of all the amino acids. Mammalian MTs contain a total of seven equivalents of zinc and/or cadmium, and sometimes also copper, as Cu(I), as well as Hg(II), Ag(I), Au(I) and Bi(III). The function of mammalian MT is unknown but it appears to account for the accumulation of cadmium in the body and may be involved in its detoxification. Cadmium can apparently regulate the rate of MT gene transcription (Kägi and Kojima, 1987).

Based on the position of cysteine residues in their amino acid sequences, the MTs of sea urchin, *Saccharomyces cerevisiae* and the cyanobacterium *Synechococcus* TX-20 resemble those of mammals, whereas the metal thiolate polypeptides of fission yeast and plants which are poly(γ-glutamylcysteinyl)glycines (Kägi and Kojima, 1987) are more distant.

The $Cd^{2+}$ ion has a similar ionic radius to $Ca^{2+}$ (0.95 Å) and because of its position in the Periodic Table (Group IIB) also bears a chemical resemblance to $Zn^{2+}$. It will therefore bind at the predominantly oxygen-containing ligand sites (glutamate, aspartate) preferred by $Ca^{2+}$, and sites normally occupied by $Zn^{2+}$ containing a sulphur (cysteine, as in MT), nitrogen (histidine) and oxygen ligands. It is notable that $Cd^{2+}$ has been reported to *enhance* the activities of many enzymes, including some which are zinc-dependent (Vallee, 1979). The role of cadmium proteins in bacterial growth is, however, not clear.

*Euglena gracilis* provides an example of antagonism between $Cd^{2+}$ and $Zn^{2+}$: cells growing in zinc-rich media do not take up $Cd^{2+}$ readily (Falchuk *et al.*, 1975). Similarly, Laddaga and Silver (1985) have shown that $Zn^{2+}$ is a competitive inhibitor of $Cd^{2+}$ uptake by *Escherichia coli*.

## Cadmium uptake by *Pseudomonas putida*

*Growth and adaptation*

We studied a strain of *Pseudomonas putida* isolated from sewage water which adapted to growth in a defined chemical medium (amino acids, glucose, β-glycerophosphate, a few mineral salts and Tris buffer) in the presence of increasing levels of $Cd^{2+}$ (Higham *et al.*, 1985, 1986). Adaptation involved increases in growth rate, a reduction in the length of the lag phase and reduction in the extent of cadmium accumulation from the medium. A lag phase of up to 7 h was observed when adapted cells were transferred to a fresh medium containing 3 mM $Cd^{2+}$. Small amounts of $Zn^{2+}$ (60 $\mu$M) in the medium decreased the lag phase, increased the growth rate and yield of cells. Adapted cells grown in the absence of cadmium quickly lost their resistance. Cells actively concentrated $Cd^{2+}$ in two distinct kinetic phases, and tolerated high intracellular cadmium concentrations ($>6$ mM). There appeared to be mechanisms by which cells could expel cadmium during the lag and exponential phases, and resistant cells adapted so as to control both the extent and the rate of cadmium uptake as compared to control cells.

## Membrane structure

During the long lag phase in a medium containing 3 mM $Cd^{2+}$, cadmium-adapted *P.putida* cells exhibited extensive blebbing of the outer membrane. The blebs appeared to be membrane vesicles, some of which were continuous with the outer membrane. Polyphosphate granules containing cadmium also appeared to be present in the cells. Exponential phase cells readily clustered, were smaller than control cells, and non-motile. Cadmium-adapted cells released more lipopolysaccharide into the medium than did control cells. Addition of $Ca^{2+}$, but not $Mg^{2+}$, prevented this release. Changes in membrane structure were also indicated by increasing sensitivity towards certain antibiotics including aminoglycosides, cyclic polypeptides and doxycycline. It appeared, therefore, that resistance, under the conditions we used, involved a complicated series of events and processes: including cadmium complexation to polyphosphate granules, structural changes in cell membranes and perhaps intracellular cadmium-binding proteins.

## Resistance mechanisms

Cadmium resistance mechanisms in bacteria have been reviewed (Silver and Misra, 1988; Trevors *et al.*, 1985, 1986; Summers and Silver, 1978) but the possible role of intracellular cadmium-binding proteins has received relatively little attention. Best understood are the cadmium *efflux systems* of the Gram-positive bacterium *Staphylococcus aureus*. $Cd^{2+}$ appears to enter via the $Mn^{2+}$ transport system and is rapidly effluxed from resistant cells via an antiporter exchanging $Cd^{2+}$ for $2H^+$ (Tynecka *et al.*, 1981a,b) and cation-translocating ATPase (Chapter 4). Reduced cadmium accumulation has also been reported for *Bacillus subtilis* (Surowitz *et al.*, 1984; Laddaga *et al.*, 1985).

Resistance mechanisms involving cadmium *precipitation* as the sulphide or phosphate have been described for *Klebsiella aerogenes* (Aiking *et al.*, 1982, 1984) and *Citrobacter* sp. (Macaski and Dean, 1984; Macaskie *et al.*, 1987). Mitra *et al.* (1975) reported that the lag period induced in *E.coli* by 3 mM cadmium decreased with successive subcultures, and suggested that this phase involved the repair of cadmium-induced DNA single strand breaks. They also suggested the presence in *E.coli* of cadmium-binding proteins (Mitra, 1984; Khazaeli and Mitra, 1981). The best characterized prokaryotic cadmium-binding protein is that from the cyanobacterium *Synechococcus* sp. (Olafson, 1984; Olafson *et al.*, 1980) which contains $Cys-X-Cys$ repeats (where X is an amino acid other than cysteine) in its amino acid sequence, a feature found in MTs.

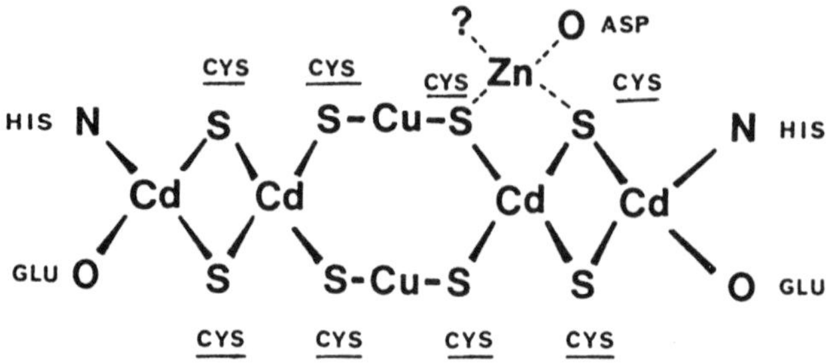

**Figure 1.** A model for the metal-binding sites of the major cadmium-containing protein isolated from cadmium-resistant *P.putida* (adapted from Higham *et al.*, 1984).

## Cadmium proteins in P.putida

Cadmium-adapted *P.putida* bacteria growing in a defined medium containing 3 mM $Cd^{2+}$ actively accumulated cadmium to an overall intracellular concentration of 4−9 mM, of which about 60% was in the cell envelope. We separated cytoplasmic cadmium-binding components using gel filtration and ion-exchange chromatography, and purified three cadmium-containing proteins (Higham *et al.*, 1984, 1986). The major protein, produced during exponential phase, had an $M_r$ of ~7200, an isoelectric point of 8.3, and contained 4.2, 0.9 and 1.8 g ions of cadmium, zinc and copper respectively per mol protein. The high cysteine content (12−23%) of all three proteins, although lower than that of mammalian MTs (33%), suggested that there may be a relation between them.

The major purified protein changed from colourless to yellow-brown in the presence of air, perhaps due to oxidation of Cu(I) to Cu(II). Based on the well-resolved cadmium-113 NMR spectrum and other data, we were able to propose a model for the metal binding sites (*Figure 1*).

After completing our initial studies on this system, we attempted to obtain another large batch of these proteins using as inoculum a sample of the original cells which had been stored freeze-dried and frozen. However, growth in the same defined medium was now very poor even in 1 mM $Cd^{2+}$ and we were unable to detect the same low $M_r$ cadmium-containing proteins in gel filtration or ion exchange chromatography profiles. Much of the cytoplasmic cadmium appeared to reside in much higher molecular weight species.

We could speculate on the reasons for such changes. These would include the lack of definition of many components of our so-called defined growth medium, e.g. manganese, iron, selenium, etc., and their subsequent variation as trace impurities in the stock chemicals used, and genetic changes including methylation of DNA bases (known to affect transcription), and the behaviour of plasmids. In preliminary work, J.T.Trevors and M.E.Starodub (University of Guelph, unpublished results) have isolated a 60 Md plasmid from our strain of *P.putida*. Horitsu *et al.* (1986) have reported the presence of a 52 kb plasmid in cadmium-resistant *P.putida* GAM-1, which transformed *E.coli* C600 to cadmium resistance by markedly reducing cadmium uptake.

An attempt was made to improve growth by using rich broth media. It is important to understand the composition of such media, since complexation of cadmium to components may greatly influence cadmium uptake into the bacteria.

42

## Broth culture media

The broth medium we examined was Luria−Bertani (LB) broth composed of tryptone (a pancreatic digest of casein milk proteins), yeast extract (the soluble supernate from autolysed yeast cells) and NaCl in a 2:1:1 w/w ratio. The organic components have been studied by $^1H$ and phosphorus-31 NMR and the elemental composition by atomic absorption, inductively-coupled plasma (ICP) spectrophotometry and ICP-mass spectrometry. Only brief details can be given here.

By $^1H$ NMR a variety of low $M_r$ components such as amino acids and sugars were detected, as expected, together with higher molecular weight species, which produced broader resonances. It was possible to monitor the uptake of these components from the medium during bacterial growth. Phosphorus-31 NMR resonances included those for phospholipids, phosphate and phosphoserines, the latter being prevalent in casein proteins.

The levels of some of the metal ions present were measured to be: $Na^+$, 92 mM; $K^+$, 5.7 mM; $Ca^{2+}$, 0.5 mM; $Mg^{2+}$, 0.2 mM; $Fe^{3+}$, 20 $\mu$M; $Zn^{2+}$, 14 $\mu$M; $Cu^{2+}$, 0.2 $\mu$M; $Mn^{2+}$, 0.1 $\mu$M; $Cd^{2+}$, 0.02 $\mu$M. Thus, even control bacteria growing in such a medium will be exposed to cadmium. By coupling fast protein liquid chromatography (FPLC) to atomic absorption and ICP we have been able to monitor the distribution of metals amongst the various $M_r$ fractions of the broth. Iron and zinc appeared to be present mainly in components of $M_r$ 2−7 k whereas added cadmium was distributed in the $M_r$ <2 k range.

### Cadmium precipitation in broth media

Addition of simple cadmium salts at levels greater than about 1 mM to LB broth resulted in precipitation. The precipitate contained high amounts of cadmium and phosphate but also organic components. We found that precipitation also involved the removal from the medium of essential trace elements such as zinc and iron. When 10 mM $CdSO_4$ was added, ∼50% of the zinc and iron also precipitated. Most of our subsequent experiments using LB broth were therefore carried out at cadmium concentrations of 1 mM or less.

### Cadmium uptake by E.coli

When *E.coli* K12 C600, a well-defined strain, was cultured in LB broth containing increasing concentrations of $Cd^{2+}$, an increase in lag phase, prolongation of the exponential phase and a decreased yield were observed (*Figure 2*). The experiments described below were carried out on non-adapted cells.

The distribution of cadmium in cytoplasmic extracts was determined by FPLC with simultaneous monitoring of the eluant for cadmium by atomic absorption, and absorbance at 255 nm (*Figure 3*). By this technique, separations can be completed in about 30 min. Cells were grown in LB broth containing 1 mM $Cd^{2+}$ for 5 days, then washed twice with LB broth and sonicated. Four well-resolved cadmium-containing peaks are seen; the two most intense ones (labelled **c, d** in *Figure 3*) correspond to molecular weight values of more than ∼50 k, and the other two (**a, b**) to $M_r$ values of less than ∼7 k. Peaks **b, c** and **d** gave rise to clearly defined, although weak, cadmium peaks when rechromatographed on a Mono Q anion exchange column and eluted with an NaCl gradient.

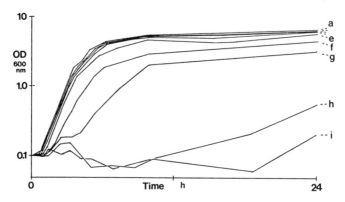

**Figure 2.** Growth curves for *E.coli* K12 C600 cells at 30°C in the presence of increasing concentrations of $Cd^{2+}$ in LB broth (1 OD unit is equivalent to $8 \times 10^8$ cells $ml^{-1}$). Key to cadmium concentrations: **a**, 0 mM; **b**, 0.01 mM; **c**, 0.022 mM; **d**, 0.046 mM; **e**, 0.10 mM; **f**, 0.22 mM; **g**, 0.46 mM; **h**, 1.0 mM; **i**, 2.2 mM. At 2.2 mM $Cd^{2+}$ about 10% of the cadmium precipitates.

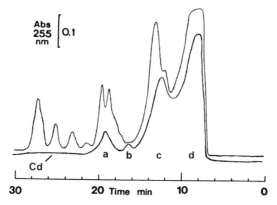

**Figure 3.** Fractionation of a cytoplasmic extract from *E.coli* grown in LB broth containing 1 mM $Cd^{2+}$ on a Superose 12 gel filtration FPLC column, as monitored by cadmium concentration (atomic absorption, lower trace) and absorbance at 255 nm (upper trace).

Addition of EDTA to extracts appeared to mobilize cadmium preferentially from fraction **c** into low $M_r$ forms (CdEDTA); similarly, peak **c** was much diminished in intensity if cells were washed with EDTA before sonication. Peak **c** was also prominent, whereas peak **d** was weak, when $Cd^{2+}$ was added to cytoplasmic extracts from control cells. This raises the question of the possible redistribution of $Cd^{2+}$ from the cell-wall into the cytoplasm during the sonication and fractionation procedures. However, washing experiments appear to show that cells grown in cadmium-containing media contained more tightly-bound cadmium than cells incubated in cadmium-containing media at 4°C (*Figure 4*). In the latter case, most of the cadmium is probably in the cell-walls and can be washed out almost completely with LB broth containing EDTA. EDTA is known to lead to extraction of lipopolysaccharide from *E.coli* (Beveridge and Koval, 1981; Chapter 5). In contrast, cells grown in the presence of cadmium retained a considerable amount of cadmium after similar washing procedures (*Figure 4*) suggesting that the metal is tightly bound in the cytoplasm.

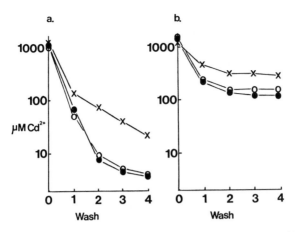

**Figure 4.** Graphs showing the total concentration of cadmium in suspensions of control *E. coli* cells (**a**) and cells grown in 1 mM $Cd^{2+}$ (**b**), resuspended in LB broth containing 1 mM $Cd^{2+}$ at 4°C. For the first wash, cells were incubated with LB broth ($\times$), LB broth + 10 mM EDTA ($\bigcirc$) or LB broth + 100 mM EDTA ($\bullet$), and subsequent washes were with LB broth only.

## Conclusions

A variety of resistance mechanisms enable bacteria to adapt to growth in media containing high concentrations of cadmium. These mechanisms include extracellular precipitation as the sulphide or phosphate, and efficient pumping out if cadmium enters the cell. The latter mechanism, which appears to be accomplished by membrane-bound proteins, is available because $Cd^{2+}$ is a relatively labile cation and undergoes rapid ligand exchange reactions; for $H_2O$ ligands, exchange rates follow the order $Cd^{2+} > Ca^{2+} > Zn^{2+}$. This kinetic feature may allow discrimination between $Cd^{2+}$ and $Zn^{2+}$ or $Ca^{2+}$, and may be enhanced by the ability of $Cd^{2+}$ to bind to both hard (oxygen and nitrogen) and soft (sulphur) ligands, and its coordination number of four to six. It is not clear how $Cd^{2+}$ interferes with the biochemistry of $Mn^{2+}$ in some bacteria (Laddaga and Silver, 1985). $Mn^{2+}$ biochemistry resembles that of $Mg^{2+}$ and involves oxygen and some nitrogen binding sites.

The efflux mechanism is probably not sufficient when cells are subjected to very high (millimolar) levels of cadmium. Cadmium entering the cytoplasm can be chelated by polyphosphate during early growth stages, and in later stages binding proteins appear to play a role. Cadmium proteins related to the cysteine-rich MTs have been isolated from *P. putida* and cyanobacteria, but the conditions under which they are produced remain to be clarified. The difficulties associated with the isolation of MTs from bacteria may simply be due to lack of stability during isolation since they are prone to oxidation especially if the copper content is high. Rapid separation methods, such as those described here, based on HPLC in glass apparatus combined with on-line metal detection, should facilitate future work. The problems of disrupting bacteria so as to isolate cytoplasmic components without allowing migration of cadmium from cell-wall compartments also requires further attention. It is possible that small cadmium-containing proteins remain bound to DNA and RNA during the isolation procedures. We have found that both *P. putida* and *E. coli* cytoplasmic extracts contain a considerable

45

proportion of cadmium in the very high $M_r$ fraction.

The choice of an appropriate growth medium for studies of bacterial adaptation to cadmium is a problem. Defined chemical media are attractive because the state of cadmium complexation in them can be determined. However, knowledge of which elements are essential for growth is probably not complete for certain bacteria and the trace element content of common laboratory chemicals such as amino acids and buffers is relatively high. It is therefore sometimes difficult to achieve reproducible growth conditions. Growth yields of bacteria grown under high cadmium stress in culture are often so poor that broth media are preferred. Here the state of cadmium complexation is uncertain because of the ill-defined nature of many of the components of the medium. It is important to explore methods which enable both the composition of broth media and metal complexation in them to be determined. We have shown that $^1H$ and phosphorus-31 NMR studies provide a useful approach to these problems but others will also be needed.

## Acknowledgements

We thank Dr Michael Scawen (Porton) for stimulating discussions and collaboration during the course of this work, and are grateful to the SERC, Public Health Laboratories and MRC for support.

## References

Aiking,H., Kok,K., van Heerikhuizen,H. and van 'T Riet,J. (1982) Adaptation to cadmium by *Klebsiella aerogenes* growing in continuous culture proceeds mainly via formation of cadmium sulfide. *Applied and Environmental Microbiology* **44**, 938−944.

Aiking,H., Stijnman,A., van Garderen,C., van Heerikhuizen,H. and van 'T Riet,J. (1984) Inorganic phosphate accumulation and cadmium detoxification in *Klebsiella aerogenes* NCTC 418 growing in continuous culture. *Applied and Environmental Microbiology* **47**, 374−377.

Bernhard,M., Brinckman,F. and Sadler,P.J. (1986) *The Importance of Chemical 'Speciation' in Environmental Processes*. Dahlem Konferenzen Life Sciences Research Report **33**, Springer-Verlag, Berlin.

Beveridge,T.J. and Kovak,S.F. (1981) Binding of metals to cell envelopes of *Escherichia coli* K12. *Applied and Environmental Microbiology* **42**, 325−335.

Birch,N.J. and Sadler,P.J. (1979) Inorganic elements in biology and medicine. In *Inorganic Biochemistry* (ed. H.A.O.Hill), Specialist Periodical Report, Vol. 1, The Chemical Society, London, pp. 356−420.

Chen,R.W. and Ganther,H.E. (1975) Cadmium metallothionein. *Environmental Physiology and Biochemistry* **5**, 378−388.

Falchuk,K.H., Fawcett,D.W. and Vallee,B.L. (1975) Competitive antagonism of cadmium and zinc in the morphology and cell division of *Euglena gracilis*. *Journal of Submicroscopic Cytology* **7**, 139−152.

Higham,D.P., Sadler,P.J. and Scawen,M.D. (1984) Cadmium-resistant *Pseudomonas putida* synthesizes novel cadmium proteins. *Science* **225**, 1043−1046.

Higham,D.P., Sadler,P.J and Scawen,M.D. (1985) Cadmium resistance in *Pseudomonas putida*: growth and uptake of cadmium. *Journal of General Microbiology* **131**, 2539−2544.

Higham,D.P., Sadler,P.J. and Scawen,M.D. (1986) Effect of cadmium on the morphology, membrane integrity and permeability of *Pseudomonas putida*. *Journal of General Microbiology* **132**, 1475−1482.

Higham,D.P., Sadler,P.J. and Scawen,M.D. (1986) Cadmium-binding proteins in *Pseudomonas putida*: pseudothioneins. *Environmental Health Perspectives* **65**, 5−11.

Horitsu,H., Yamamoto,K., Wachi,S., Kawai,K. and Fukuchi,A. (1986) Plasmid-determined cadmium resistance in *Pseudomonas putida* GAM-1 isolated from soil. *Journal of Bacteriology* **165**, 334−335.

Kägi,J.H.R. and Kojima,Y. (eds.) (1987) *Metallothionein* II. Birkhauser Verlag, Basel.

Khazaeli,M.B. and Mitra,R.S. (1981) Cadmium-binding component in *Escherichia coli* during accommodation to low levels of this ion. *Applied and Environmental Microbiology* **41**, 46−50.

Laddaga,R.A., Bessen,R. and Silver,S. (1985) Cadmium-resistant mutant of *Bacillus subtilis* 168 with reduced cadmium transport. *Journal of Bacteriology* **162**, 1106−1110.

Laddaga,R.A. and Silver,S. (1985) Cadmium uptake in *Escherichia coli* K-12. *Journal of Bacteriology* **162**, 1100−1105.

Macaskie,L.E. and Dean,A.C.R. (1984) Cadmium accumulation by a *Citrobacter* sp. *Journal of General Microbiology* **130**, 53−62.

Macaskie,L.E., Dean,A.C.R., Cheetham,A.K., Jakeman,R.J.B. and Skarnulis,A.J. (1987) Cadmium accumulation by a *Citrobacter* sp.: the chemical nature of the accumulated metal precipitate and its location on the bacterial cells. *Journal of General Microbiology* **133**, 539−544.

Mitra,R.S., Gray,R.H., Chin,B. and Berstein,I.A. (1975) Molecular mechanisms of accommodation in *Escherichia coli* to toxic levels of cadmium. *Journal of Bacteriology* **121**, 1180−1188.

Mitra,R.S. (1984) Protein synthesis in *Escherichia coli* during recovery from exposure to low levels of cadmium. *Applied and Environmental Microbiology* **47**, 1012−1016.

Olafson,R.W. (1984) Prokaryotic metallothionein. *International Journal of Peptide and Protein Research* **24**, 303−308.

Olafson,R.W., Loya,S. and Sim,R.G. (1980) Physiological parameters of prokaryotic metallothionein induction. *Biochemical and Biophysical Research Communications* **95**, 1495−1503.

Sadler,P.J., Higham,D.P. and Nicholson,J.K. (1985) The environmental chemistry of metals with examples from studies of the speciation of cadmium. In *Environmental Inorganic Chemistry* (eds K.J.Irgolic and A.E.Martell), VCH Publishers, Deerfield Beach, FL, pp. 249−272.

Schwarz,K. (1977) Essentiality versus toxicity of metals. In *Clinical Chemistry and Toxicology of Metals* (ed. S.S.Brown), Elsevier, Amsterdam, pp. 3−22.

Schwarz,K. and Spallholz,J. (1978) Growth effects of small cadmium supplements in rats maintained under trace element controlled conditions. *Proceedings of the First International Cadmium Conference, 1977*, Metals Bulletin Ltd, Worcester Park, pp. 105−109.

Sherlock,J.C. (1983) The uptake by man of cadmium from vegetables grown on sludged land. *Proceedings of the Fourth International Cadmium Conference*, Munich, Cadmium Association, London, pp. 107−109.

Silver,S. and Misra,T.K. (1988) Plasmid-mediated heavy metal resistances. *Annual Review of Microbiology* **42**, 717−743.

Summers,A.O. and Silver,S. (1978) Microbial transformations of metals. *Annual Review of Microbiology* **32**, 637−672.

Surowitz,K.G., Titus,J.A. and Pfister,R.M. (1984) Effects of cadmium acccumulation on growth and respiration of a cadmium-sensitive strain of *Bacillus subtilis* and a selected cadmium resistant mutant. *Archives of Microbiology* **140**, 107−112.

Trevors,J.T., Oddie,K.M. and Belliveau,B.H. (1985) Metal resistance in bacteria. *FEMS Microbiology Reviews* **32**, 39−45.

Trevors,J.T., Stratton,G.W. and Gadd,G.M. (1986) Cadmium transport, resistance, and toxicity in bacteria, algae, and fungi. *Canadian Journal of Microbiology* **32**, 447−464.

Tynecka,Z., Gos,Z. and Zajac,J. (1981a) Reduced cadmium transport determined by a resistance plasmid in *Staphylococcus aureus*. *Journal of Bacteriology* **147**, 305−312.

Tynecka,Z., Gos,Z. and Zajac,J. (1981b) Energy-dependent efflux of cadmium coded by a plasmid resistance determinant in *Staphylococcus aureus*. *Journal of Bacteriology* **147**, 313−319.

Vallee,V.L. (1979) Metallothionein: historical review and perspectives. In *Metallothionein* (eds J.H.R.Kagi and M.Nordberg), Birkhauser Verlag, Basel, pp. 19−40.

CHAPTER 4

# Plasmid-determined resistance to metal ions

SIMON SILVER[1], RICHARD A.LADDAGA[2] and TAPAN K.MISRA[1]

[1]*University of Illinois College of Medicine, Chicago, IL 60680, USA and* [2]*Bowling Green State University, Bowling Green, OH 43405, USA*

## Introduction

Bacterial cells have co-existed with toxic heavy metals since the origin of life, perhaps 3 or 4 × $10^9$ years ago. In the early stages of the evolution of life, when volcanic activity and other sources of toxic heavy metals were ubiquitous, it was essential to invent mechanisms to cope with the toxic heavy metals that were abundant in the environment. Perhaps later in prokaryotic evolution, the genes for resistances to toxic heavy metals were less essential, and it became important to minimize the genetic burden of carrying these sometimes essential, sometimes burdensome genes. Then, as best as we can reconstruct, the genes for toxic heavy metal resistances, quite similarly to the genes for antibiotic resistances and other ancillary functions, were 'packaged' into bacterial plasmids and transposons. These small circular pieces of DNA and linear 'hopping genes' respectively could be lost from particular lines of bacterial cells easily, without endangering the rest of the cellular genetic heritage; even more importantly, they could move from place to place, from cell to cell, in a facile fashion by mechanisms of intercellular gene transfer, thus facilitating the spread of a resistance among a population of bacteria including more than a single strain or species, when stress from toxic heavy metal pollution made this resistance an asset for cellular survival.

With this view of how bacterial toxic heavy metal resistances came about, we can recognize that they should be quite ancient, and we can expect that the genetic and biochemical mechanisms involved will share properties with those for essential cellular roles, such as growth, metabolism of energy and carbon sources, and biosynthesis of essential nutrients.

In common with resistances to antibotics, bacterial resistances to toxic heavy metals appear to have only a small number of basic mechanisms.

(i)     There are enzymes, oxidases and reductases, which convert metal ions from more toxic into less toxic. There are alkylating enzymes and de-alkylating lyases, that add and remove covalently attached components of organometal compounds. Reduction and dealkylation are the major mechanisms of resistance to inorganic and organic mercury compounds. Arsenite oxidation and chromate reduction can be mechanisms of resistance for those toxic metal ions, but they have not been demonstrated (as yet) to be plasmid resistance mechanisms.

(ii)    There is the possibility of sequestration and binding of toxic heavy metals either in the cell-wall (preventing them from reaching the intracellular cytoplasm) or

49

intracellularly in highly specific binding components, such as metallothionein. Again, there is no clear-cut example where this has been demonstrated to be a plasmid gene-determined resistance mechanism.

(iii)     There is the possibility of blocking cellular uptake by altering the uptake pathway available in sensitive cells. Although this is the mechanism of chromosomally-determined resistance to arsenate, cadmium, chromate and perhaps other heavy metal ions, there is no known example of this mechanism for plasmid-governed resistance.

(iv)     Once the toxic heavy metal ion has reached the intracellular cytoplasm, it can be pumped out again rapidly by a highly specific efflux system, which might derive its energy either from the membrane potential or more directly from ATP. Efflux pumps seem to be the mechanism of resistance to tetracycline (an antibiotic) in bacteria and for resistance to arsenic, cadmium and chromate, at least in some cases. In the last few years, DNA sequencing analysis has clarified the nature of an arsenic ATPase and a cadmium ATPase efflux system, both of which are highly specific.

(v)     There is the possibility of altering intracellular targets for the toxic action of heavy metals, but in the case of metal ions, rather than antibiotics, the mechanisms of toxicity are generally so broad and non-specific that this might be impossible. It might be necessary to change all of the thiol-containing enzymes that are sensitive to cadmium or arsenite in order to obtain resistance to these toxic ions. In terms of genetically-determined biochemical functions, that is not feasible.

This brief overview will emphasize recent findings from DNA sequence analysis that has provided detailed understanding for cadmium, arsenic and mercury resistance governed by bacterial plasmids. A more detailed coverage of bacterial heavy metal resistance appeared recently (Silver and Misra, 1988). This summary of recent progress in DNA sequencing and molecular analysis of plasmid-determined heavy metal resistance mechanisms is a revised version of a report from another symposium that will appear in *Biological Trace Element Research* (1989).

## Cadmium resistance

There are perhaps six known mechanisms of cadmium resistance (Silver and Misra, 1988) and only one of these has been studied and understood at a reasonable level from DNA sequencing analysis. This is the *cadA* determinant from *Staphylococcus aureus* plasmids (Novick *et al.*, 1979). Weiss *et al.* (1978) showed that cadmium accumulation occurs via the chromosomally-determined manganese transport system in sensitive cells and that less net cadmium accumulation occurs in resistant cells. Rather than a direct block on $Cd^{2+}$ uptake, Tynecka *et al.* (1981) demonstrated that an energy-dependent efflux system functions in resistant cells but is missing in sensitive cells. Based on studies of inhibition of cadmium efflux by membrane-perturbing antibiotics, Tynecka *et al.* (1981) concluded that the fundamental mechanism was a $Cd^{2+}/2H^+$ exchange (*Figure 1*). It remained to be seen whether the efflux system was powered by the pH gradient across the cellular membrane or directly coupled to ATP energy. Recent DNA sequence analysis of the *cadA* determinant has clarified this important aspect (Nucifora *et al.*, 1989a; Silver *et al.*, 1989). The cadmium resistance determinant contains only a single,

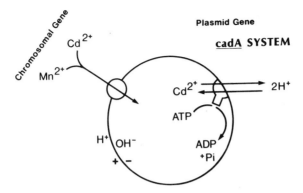

**Figure 1.** Cadmium uptake by the manganese transport system and efflux by the *cadA* $Cd^{2+}$ efflux ATPase. (Modified from Silver and Misra, 1988.)

but very long, open reading frame. The *cadA* gene potentially encodes a polypeptide of 727 amino acids. It is surprising that a protein whose function is to confer cadmium resistance is unusually low in cysteine content, with only 4 out of 727 residues. However, these residues are strategically positioned.

The comparison of the *cadA* amino acid sequence with other polypeptide sequences in DNA-derived polypeptide sequence libraries showed that the $Cd^{2+}$ resistance protein was a member of a growing family of cation-translocating E1, E2 ATPases, that includes the previously determined $K^+$ uptake ATPases of both Gram-positive (Solioz *et al.*, 1987) and Gram-negative (Walderhaug *et al.*, 1987) bacteria, the $H^+$ efflux ATPases of yeast and *Neurospora*, and even the $Ca^{2+}$ ATPase of mammalian muscle and the $Na^+/K^+$ ATPases of animal cell membranes (Nucifora *et al.*, 1989a; Silver *et al.*, 1989; Walderhaug *et al.*, 1987). The E1, E2 ATPases differ from the $F_1$, $F_0$ class of ATPases in that they have a phosphorylated intermediate state and the two states of the enzyme, E1 and E2, can be isolated and studied. The basic properties of these enzymes, deduced from 25 years of direct biochemistry on the $Ca^{2+}$ and $Na^+/K^+$ ATPases plus the close sequence analogies with the other E1, E2 ATPases, are shown in *Figure 2* (Silver *et al.*, 1989). This model can be considered no more than a cartoon at the moment. The basic properties of this protein seem clear, although it has yet to be directly measured. The model should be taken cautiously. The CadA polypeptide starts with an amino-terminal substrate recognition sequence. The first two cysteines at positions 23 and 26 are closely homologous in sequence to $Hg^{2+}$ binding regions hypothesized in the mercuric reductase and mercury-binding protein that are components of the mercury resistance system (see below). A recurring pattern seems to be developing with a soft metal dithiol binding motif, now in two of these resistance systems. Following a pair of transmembrane segments (perhaps from residues 106–126 and 130–150), there is a domain of approximately 186 residues on the cytoplasmic side of the membrane that constitutes a cation 'funnel' accepting the $Cd^{2+}$ cation from the cysteine pair and guiding it to the transmembrane channel. In the case of the $Ca^{2+}$ ATPase, there is direct data for this (summarized in Brandl *et al.*, 1986). For the new $Cd^{2+}$ ATPase it is entirely hypothetical, derived from DNA sequence analysis. The second pair of

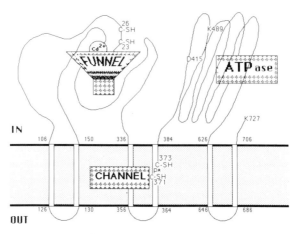

**Figure 2.** Hypothetical model of the $Cd^{2+}$ ATPase from the amino acid sequence and homology to E1, E2 ATPases (see text). (Modified from Silver *et al.*, 1989.)

transmembrane, potentially $\alpha$-helical segments from approximately 336 to 356 and 364 to 384 are positioned in a comparable setting to the transmembrane components thought to constitute the 'channel' itself in the other E1, E2 ATPases. The second cysteine pair occurs here at positions 371 and 373, bounding a proline residue (*Figure 2*) that occurs in a comparable position in all of these enzymes (Silver *et al.*, 1989). Then follows an ATPase domain of approximately 250 amino acids in length, including an aspartate residue at position 415, which starts a sequence of seven amino acids unaltered in all of these proteins (Silver *et al.*, 1989). Lysine$_{489}$ is a candidate for binding ATP (by analogy to the $Ca^{2+}$ and $Na^+/K^+$ enzymes) prior to donating the phosphate to aspartate$_{415}$. The end of the ATPase domain and the next potential membrane spanning segment are the most conserved sequences in all of this family of proteins (Brandl *et al.*, 1986; Silver *et al.*, 1988). Following a third pair of membrane spanning potentially $\alpha$-helical segments, the protein ends on the cytoplasmic side with lysine$_{727}$ (*Figure 2*). This general pattern of (i) substrate recognition domain, followed by (ii) a pair of closely positioned membrane spanning segments, (iii) a substrate sequestering funnel on the cytoplasmic side, (iv) another pair of transmembrane peptides, potentially lining the cation-specific transmembrane channel, (v) a large highly conserved ATPase domain, followed by (vi) a third pair of membrane spanning segments is common to all of these proteins. The model in *Figure 2* is similar to those for the better studied $Ca^{2+}$ and $Na^+/K^+$ ATPases in that the bulk of the amino acid residues (perhaps 75%) are cytoplasmic, with perhaps 20% of the residues being in the membrane in this integral membrane protein; only a very small portion of the polypeptide is extracellular. This pattern should be subject to limited proteolysis, as has been done with the $Ca^{2+}$ and $Na^+/K^+$ ATPases. Even in a report on DNA sequence analysis of resistance mechanisms, the reader should be warned that the model in *Figure 2* (although detailed in its structural and functional specifics) is based entirely on the DNA sequence analysis, without direct biochemical data at the moment. For a system for which the DNA sequence has not as yet been published, the supporting experimental efforts have not been undertaken.

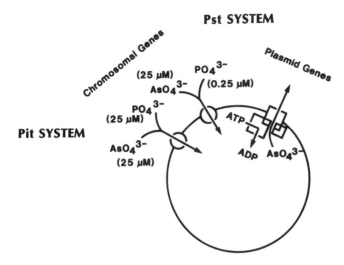

**Figure 3.** Arsenate uptake by the phosphate transport systems of *E.coli* and arsenate efflux by the plasmid-determined arsenic ATPase. (From Silver and Misra, 1988.)

## Arsenic resistance

The arsenic resistance ATPase of Gram-negative bacteria has been more thoroughly studied at the biochemical level (Chen *et al.*, 1986; Silver and Misra, 1988; Rosen *et al.*, 1988), following earlier physiological studies (Silver *et al.*, 1981) and then DNA sequence analysis (Chen *et al.*, 1986). A plasmid-determined resistance system for As(III), As(V) and Sb(III) is found on plasmids of both Gram-negative and Gram-positive bacteria. Silver *et al.* (1981) showed that arsenate enters the bacterial cells via the chromosomally-determined phosphate transport systems (*Figure 3*), of which for *Escherichia coli* (and perhaps for many other bacteria) there are two. As summarized in *Figure 2*, the Pit (*Pi transport*) system has low specificity, with a $K_m$ of 25 $\mu$M phosphate and an equivalent $K_i$ of 25 $\mu$M arsenate. The Pst (*phosphate specific transport*) system has a 100 times higher affinity for phosphate than for arsenate. Arsenate, however, enters both systems with kinetic constants apparently unchanged by the plasmid resistance system (Silver *et al.*, 1981). With whole-cell experiments (Silver and Keach, 1982; Mobley and Rosen, 1982), it was demonstrated that efflux of accumulated arsenate was an energy-dependent process with characteristics indicative of an ATPase mechanism. The DNA sequence analysis indicated the existence of four genes. The first (*arsR*) determines a regulatory protein which binds to the DNA, switching on the arsenic resistance system when induced (Silver *et al.*, 1981). The DNA sequence of the *arsR* gene is complete but not published (B.P.Rosen, personal communication). The DNA sequence analysis (Chen *et al.*, 1986) of the other three genes (*arsA, arsB* and *arsC*) has shed significant light upon the mechanism of arsenic efflux, but not allowed the detailed understanding (by homology to other systems) that was true for the cadmium ATPase. This is because the arsenic ATPase is less similar to other known ATPases than are the $Cd^{2+}$ ATPase and other E1, E2 ATPases.

The ArsA protein is a loosely membrane-associated protein that has been purified to homogeneity and shown to function *in vitro* as an arsenic-stimulated ATPase (Rosen *et al.*, 1988). From the amino acid sequence, this polypeptide has two potential ATP-binding regions, homologous to those found in other ATPases of both bacterial and eukaryotic origin (Chen *et al.*, 1986). The homologies of these ATP binding sequences have recently been discussed in more detail (Silver *et al.*, 1989). The ArsB protein potentially encodes a very high hydrophobic, probably integral, membrane protein. From its sequence and from computer modelling ArsB is thought to go back and forth across the membrane as many as twelve times (Chen *et al.*, 1986; Silver *et al.*, 1989). ArsB potentially contains the arsenic and antimony transmembrane channel, but these residues have not yet been identified. ArsA and ArsB are essential for the working of the arsenic/antimony efflux pump. Both are large polypeptides, of approximately 500 amino acids in length. In comparison, the ArsC polypeptide is smaller, only 141 amino acids in length, and it appears to determine the substrate specificity of the efflux ATPase rather than being essential for its functioning. When the *arsC* gene was removed by deletion from the end, the system still functioned and afforded arsenite resistance, but arsenate resistance was lost (Chen *et al.*, 1986).

Comparison of the amino acid sequences of the ArsA and ArsC polypeptides (from the DNA sequences) with those of other recently published polypeptide sequences led B.P.Rosen (personal communication) to a tentative hypothesis as to the origin of the *ars* operon. The ArsA and ArsC polypeptides are related to components of the bacterial nitrogenase system, a complex system of at least 17 gene products involved in atmospheric nitrogen fixation and a system that is rather conserved among otherwise dissimilar nitrogen-fixing bacteria. The ArsA polypeptide has two recognizable ATP-binding regions that are homologous with the *nifH*-determined iron-containing nitrogenase structural component (of which there are 12 from different bacteria in our currently available polypeptide sequence library). This ArsA/NifH homology is closer for the amino-half ATP-binding sequence of ArsA than for the carboxyl-half ATP-binding sequence of ArsA (Chen *et al.*, 1986 and Silver *et al.*, 1989 show the actual matches) but NifH is the best match in the library of ATP-binding sites for both halves of ArsA. B.P.Rosen (personal communiction) thinks it significant that ArsA is a double fusion polypeptide with similar N- and C- half sequences (undoubtedly arising from DNA duplication and fusion), whereas NifH is found functionally as a homodimeric protein.

Nitrogenase requires oxyanions such as vanadate and molybdate for function but whether this has any ancestral relationship to the oxyanion (arsenite, arsenate and antimonate) recognition sites of ArsA (Silver *et al.*, 1981; Rosen *et al.*, 1988; Silver and Misra, 1988) is not known. The ArsB membrane polypeptide has no recognizable homology to other known amino acid sequences to date. However, the small soluble ArsC sequence is significantly related (25% identities over 96 amino acid aligned segments) with a new *nif* open reading frame (ORF3) from *Azotobacter vinelandii* (Joerger and Bishop, 1988). Although it is clearly in a cluster of *nif* genes, this new open reading frame is not homologous to any previously known *nif* genes and its function is still unknown (Joerger and Bishop, 1988).

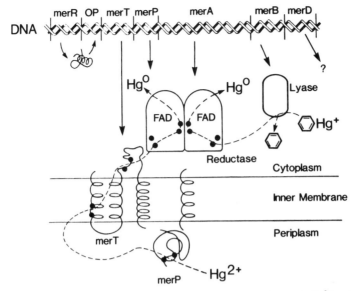

**Figure 4.** Model of the genetic determination of the system for detoxifying inorganic $Hg^{2+}$ and organomercurials (from Silver and Misra, 1988). Top line shows order of the genes on the DNA (see text). The remainder of the figure shows the protein products of the genes and their known or postulated locations and functions.

## Mercury resistance

The mercury resistance systems from a wide variety of bacterial sources, both Gram-positive and Gram-negative, have been studied in considerable detail (Summers, 1986; Foster, 1987; Silver and Misra, 1988). At the level of DNA sequencing, six have been sequenced to date, whereas there is only a single published sequence for the arsenic resistance system and the first cadmium resistance sequence is in preparation. Thus, DNA sequence analysis has provided the greatest impact to date on understanding the mercury resistance system. From studies of mutants, and cloning as well as DNA sequence analysis, the model shown in *Figure 4* has been deduced from mercury systems of Gram-negative bacteria. There is a large (more than 3 kb) segment of DNA given over to the determination of perhaps half-a-dozen proteins, which are involved in mercury resistance. Not all of the proteins occur in all of the systems studied. Starting from the left in *Figure 4*, the first gene, *merR*, produces a transacting regulatory protein. This is followed by an operator/promoter (OP) site for the association of the MerR protein to the DNA and for RNA polymerase attachment, initiating mRNA transcription. Then follows a series of structural genes, determining components of the resistance system. For the three systems from Gram-negative bacteria studied to date (Misra *et al.*, 1984, 1985; Brown *et al.*, 1986; Griffin *et al.*, 1987; Nucifora *et al.*, 1989b), these genes are *merT, merP, merC* (present only in R100, Misra *et al.*, 1984, and not in the other two systems), *merA* (the determinant of the mercuric reductase enzyme), *merB* (the determinant of the organomercurial lyase, present only in the pDU1358 system, Griffin *et al.*, 1987, of the three Gram-negative systems sequenced to date) and *merD*. From the DNA sequence analysis, followed by direct studies on the proteins and their functions (in some cases), a very sophisticated understanding is emerging. The MerR protein of Gram-negative mercuric resistant systems is a small (144 amino acids) dimeric protein that is transcribed from a gene in the opposite (right to left) orientation (Foster and Brown, 1985) from the remainder of the genes shown in *Figure 4* (transcribed left to right). The MerR protein binds to the O/P region (O'Halloran and Walsh, 1987), repressing transcription of *merR* itself and the synthesis of mRNA for the remaining genes (Foster and Brown, 1985; O'Halloran, 1989). In the absence of the protein, MerR is synthesized, and a low level of structural gene products as well (Ni'Bhriain *et al.*, 1983; Foster and Brown, 1985; O'Halloran, 1989). When MerR is bound to the DNA and $Hg^{2+}$ is added, transcription of the structural genes occurs at a high rate (O'Halloran, 1989) and the proteins are synthesized (Jackson and Summers, 1982).

The MerR protein appears to have at least four determinants encoded in its small structure. (i) There is a DNA-binding region, which appears to be a member of the 'helix-turn-helix' motif common to many DNA-binding regulatory proteins (Pabo and Sauer, 1984). (ii) There are amino acid residues, probably including four cysteines per dimer (O'Halloran, 1989) that bind $Hg^{2+}$ turning the system on. (iii) There is an organomercurial-binding region apparently at the carboxyl end of the pDU1358 version of this protein (which responds to phenylmercury as well as inorganic mercury) and which can be eliminated leaving the inorganic mercury response intact (Nucifora *et al.*, 1989b) and then (iv) there must be amino acids determining the specific interactions between the monomers forming the dimeric structure.

In all three versions of the Gram-negative mercuric resistance system, the first structural gene is *merT*, which determines a 116 amino acid membrane transport protein (Misra *et al.*, 1984; Nucifora *et al.*, 1989b). The protein (from its amino acid sequence) may pass across the cell membrane three times, as shown in *Figure 4*. The MerT protein contains two cysteine pairs, the first of which may be in the first membrane 'pass' as shown. The second cysteine pair appears to be located on the inner surface of the membrane between the second and third transmembrane segment. Deletion analysis (Lund and Brown, 1987) shows that this segment of the system determines a mercury uptake system, that in the absence of the detoxifying enzyme results in hyperaccumulation and hypersensitivity to $Hg^{2+}$ (Nakahara *et al.*, 1979).

The next gene, *merP*, determines a small periplasmic mercury binding protein that contains only 91 amino acids. As shown in *Figure 4*, this protein appears to be processed, leaving a 19 amino acid 'leader sequence' and resulting in a 72 amino acid periplasmic binding protein (Summers, 1986), which contains a single pair of cysteine residues. This mercury binding protein shows close sequence homology to the N-terminal end of the *cadA* ATPase (above) and to the N-terminal domain found in most versions of the *merA*-determined mercuric reductase enzyme (next).

The mercuric reductase protein is determined by the next gene *merA*. This protein is highly conserved (80 – 90 % amino acid identities) in the three Gram-negative versions sequenced (*Table 1*; Nucifora *et al.*, 1989b) and also quite conserved (about 40 % identities) when comparing the Gram-negative versions with the Gram-positive versions (*Table 1*). Although the Gram-negative and Gram-positive versions of the enzyme only share 40 % of their amino acid residues, when those residues known to play functional

**Table 1.** Mercuric reductase sequence homologies (percent identical amino acids) of aligned polypeptide sequences translated for three glutathione reductases, lipoamide dehydrogenase and mercuric reductase (see text and Silver and Misra, 1988 for details).

| | Percent identities | | | | | | | | Length (amino acids) |
|---|---|---|---|---|---|---|---|---|---|
| Glutathione reductase (*Pseudomonas*) | 100 | 44.7 | 44.2 | 28.9 | 30.7 | 31.0 | 32.7 | 31.8 | 451 |
| Glutathione reductase (*E.coli*) | | 100 | 54.8 | 27.7 | 31.5 | 30.0 | 31.5 | 31.5 | 450 |
| Glutathione reductase (man) | | | 100 | 28.8 | 28.5 | 28.0 | 29.4 | 28.4 | 478 |
| Lipoamide dehydrogenase (*E.coli*) | | | | 100 | 29.6 | 27.4 | 28.9 | 30.2 | 473 |
| Mercuric reductase (*Pseudomonas*) | | | | | 100 | 85.0 | 41.5 | 41.2 | 561 |
| Mercuric reductase (*Shigella*) | | | | | | 100 | 40.1 | 39.6 | 564 |
| Mercuric reductase (*S.aureus*) | | | | | | | 100 | 67.2 | 547 |
| Mercuric reductase (*Bacillus* sp.) | | | | | | | | 100 | 631 |

Boxes surround glutathione reductase comparisons, mercuric reductase comparisons, and comparisons of glutathione reductase with mercuric reductase. Alignments in *Tables 1–4* were made with the Feng and Doolittle (1987) multiple alignment program.

roles (the active site residues, and those involved in binding of FAD and NADPH) are considered, the conservation of amino acids increases to more than 90% (Laddaga *et al.*, 1987).

The locations of the FAD and NADPH binding regions of mercuric reductase are known from the close sequence and functional homologies between mercuric reductase and the enzymes of central cellular metabolism glutathione reductase and lipoamide dehydrogenase (Walsh *et al.*, 1988; Distefano *et al.*, 1989). For human glutathione reductase, a high resolution X-ray diffraction structure has been determined (Thieme *et al.*, 1981), so the location of the amino acid residues is well established. Mercuric reductase (in most versions of the enzyme available) starts with an 80 amino acid N-terminal segment which is closely homologous in sequence to the mercury binding protein (Silver and Misra, 1988). There then follow a series of domains that can be identified by homology to glutathione reductase. The 15 amino acid active site region is the most conserved part of this family of enzymes (Silver and Misra, 1988). The NADPH and FAD binding regions and even the interface region between the two subunits are significantly conserved. The C-terminal end of the enzyme is highly conserved, but differs from that of glutathione reductase and lipoamide dehydrogenase (Silver and Misra, 1988), indicating that this may be the site for mercury, binding and coordinating its entrance into the active site domain. This intricate model of how mercuric reductase functions is now being derived from direct enzymological studies and specific site-directed mutagenesis (Distefano *et al.*, 1989). *Table 1* summarizes the overall amino acid homologies among this family of enzymes, showing that glutathione reductases are more closely related one to another (44–55% identities) than they are to mercuric reductases (about 30%). Similarly, the mercuric reductases are more closely related one to the other than they are to glutathione reductase.

The next gene in the pDU1358 sequence is *merB* (Griffin *et al.*, 1987), the determinant of the organomercurial lyase enzyme. This polypeptide for which there are now four sequenced versions (two published, Griffin *et al.*, 1987; Laddaga *et al.*, 1987; and two in preparation; Wang *et al.*, 1989 and J.Altenbuchner, personal communication) is smaller (212–218 amino acids in length) and monomeric (Begley *et al.*, 1986a). The basic enzymatic mechanism of reaction (hydrolysis of organomercurials) (Begley *et al.*, 1986b) has been clarified. The *merB* gene is found only in the pDU1358 version of the three sequenced Gram-negative *mer* operons, but it occurs in all three currently sequenced Gram-positive *mer* operons. After *merB*, all three Gram-negative versions of the mercuric resistance determinant end with still another gene, *merD*, which appears to play a role in regulation of the mercury resistance system (Brown *et al.*, 1986). MerD shows significant sequence homology to MerR (Brown *et al.*, 1986), but the actual function of MerD has yet to be determined.

Although the three Gram-negative versions of the *mer* operon are closely homologous and as outlined in *Figure 4*, the three Gram-positive versions of this system known today are each quite different one from the other. The first version sequenced (Laddaga *et al.*, 1987) was from *S.aureus* plasmid pI258 and consists of a series of six or seven genes, starting with a gene for a MerR protein (which shows significant homology to that from the Gram-negative systems) followed by a series of genes which (from deletion analysis; M.Horwitz and T.K.Misra, unpublished) determine a mercuric transport function, analogous to that from the Gram-negative systems. However, the amino acid sequences of the Gram-positive transport region are rather dissimilar to those of the Gram-negative *merT-merP* transport system. After this series of perhaps four genes, the pI258 DNA sequence has a well-defined *merA* and a well-defined *merB* gene. As described above, the sequence identities between the Gram-positive and the Gram-negative versions of these two proteins are about 40%. For the *merA* gene product, the amino acid homologies are stronger in all of the known functional regions of that protein. For the MerB protein, where the functional domains have not been determined, the amino acid homology is stronger in the middle part of the polypeptide and much weaker at the amino- and carboxyl-ends (Griffin *et al.*, 1987; Laddaga *et al.*, 1987). The next Gram-positive *mer* sequence determined came from a soil *Bacillus* strain, and it occurs on the bacterial chromosome, rather than on a plasmid or transposon. We do not know the origin of these genes and how they became incorporated into the chromosome. Wang *et al.* (1989) have recently sequenced the *Bacillus mer* determinant and it is considerably different in structure from that of pI258. Both Gram-positive systems start with recognizably similar (about 59% amino acid identities) *merR* genes, followed by O/P DNA binding regions. The *Bacillus* and *S.aureus* MerR proteins are also significantly homologous with the three sequenced versions of the Gram-negative *merR* gene product (34−37% amino acid identities) and *merD* gene product (19−23% amino acid identities) (*Table 2*).

There then follows a series of three genes which are not closely related in the two systems and which in both cases may determine a mercury transport system. When optimally aligned, the *S.aureus* ORF5 amino acid sequence (possibly functionally equivalent to the Gram-negative MerT polypeptide) shows significant homologies (18−28% amino acid identities) with the *Bacillus* ORF2 product, with the ORF2 product of a new *Streptomyces* sequence (see next), and with the Gram-negative *merT*-gene product (*Table 3*).

Then, in both cases, there is the long *merA* gene that determines the mercuric reductase subunit. Surprisingly, the *Bacillus* version of this gene has a duplication and therefore the protein has two copies head-to-tail of the initial mercury-binding domain. The remainder of the two mercuric reductase sequences seem basically quite similar. Whereas

**Table 2.** Amino acid sequence homologies (percent identical amino acids) between the regulatory polypeptides (MerR) of the Gram-positive and Gram-negative *mer* systems and the *merD* gene products.

| | Percent identities | | | | | | | | Length (amino acids) |
|---|---|---|---|---|---|---|---|---|---|
| *S.aureus* ORF2 | 100 | 59.0 | 36.1 | 35.3 | 33.8 | 19.2 | 18.5 | 19.2 | 135 |
| *Bacillus* ORF1 | | 100 | 36.2 | 36.9 | 36.2 | 22.5 | 19.3 | 21.7 | 132 |
| Tn*501 merR* | | | 100 | 93.8 | 87.5 | 25.6 | 26.7 | 25.6 | 144 |
| R100 *merR* | | | | 100 | 87.5 | 26.5 | 26.7 | 26.5 | 144 |
| pDU1358 *merR* | | | | | 100 | 25.6 | 25.0 | 25.6 | 144 |
| Tn*501 merD* | | | | | | 100 | 80.0 | 90.9 | 121 |
| R100 *merD* | | | | | | | 100 | 79.2 | 120 |
| pDU1358 *merD* | | | | | | | | 100 | 121 |

**Table 3.** Amino acid sequence homologies (percent identical amino acids) between aligned sequences of presumed $Hg^{2+}$ transport proteins.

| | Percent identities | | | | Length (amino acids) |
|---|---|---|---|---|---|
| R100 *merT* | 100 | 18.5 | 28.1 | 19.6 | 116 |
| *S.aureus* ORF5 | | 100 | 20.4 | 22.5 | 129 |
| *Bacillus* ORF2 | | | 100 | 18.2 | 98 |
| *Streptomyces* ORF2 | | | | 100 | 100 |

**Table 4.** Amino acid sequence homologies (percent identical amino acids) between aligned organomercurial lyase sequences.

| | Percent identities | | | | Length (amino acids) |
|---|---|---|---|---|---|
| *Streptomyces* | 100 | 53.5 | 54.9 | 41.8 | 215 |
| *Bacillus* | | 100 | 73.2 | 40.7 | 218 |
| *S.aureus* | | | 100 | 41.4 | 216 |
| pDU1358 | | | | 100 | 212 |

the pI258 *merA* gene is followed immediately by *merB* gene (Laddaga *et al.*, 1987), there is a 2.5 kb gap between *merA* and *merB* in the *Bacillus* sequence (Wang *et al.*, 1989). What genetic determinants lie in this gap are unknown and how this 'broken operon' is regulated has still to be determined.

The newest of the mercuric resistance operons sequenced comes from a *Streptomyces lividans* strain (J.Altenbuchner, personal communication). It differs from all of the previously described systems in that all of the genes other than *merA* and *merB* are transcribed separately and in the opposite direction from the *merA* and *merB* genes. *merA* and *merB* are contiguous in this system (J.Altenbuchner, personal communication). When aligned with the three previously sequenced *merB* gene products, the *Streptomyces* organomercurial lyase is somewhat (but not strikingly) more similar to those from low G+C Gram-positive organisms than it is to the Gram-negative organomercurial lyase from plasmid pDU1358 (*Table 4*). All four organomercurial lyase sequences are more closely similarly in the middle than at the N or C termini; all four sequences have conserved residues for 4 Phe, 1 Trp, 1 Tyr, 4 Pro, 2 His and 3 Cys residues (including a conserved Trp−Cys−Ala−Leu−Asp−Thr−Leu heptapeptide) (data not shown). These conserved residues from DNA sequence analysis are candidates for essential functional residues that can be studied by mutagenesis analysis.

Continuing with surprises from the DNA sequence analysis, the *Streptomyces merA* reductase lacks the N-terminal mercury binding domain. Thus, we have three Gram-negative versions of this enzyme with closely related 80 amino acid binding domains, one of the three Gram-positive versions of the enzyme with a similar domain, another with two and a third with none!

From this exposition of the six mercuric resistance determinants that have been sequenced, it may be clear that similarities and differences shed light on functional roles and detailed biochemical mechanisms. We hope that additional sequences of arsenic, cadmium and other heavy metal resistance determinants will shed equal light on the biological function of these systems.

## Other heavy metal resistance systems

There are two other heavy metal resistance systems for which DNA sequences have been determined. These are the tellurium resistance determinant from one plasmid (Jobling and Ritchie, 1988) and the copper resistance determinant from another (Mellano and Cooksey, 1988). An outline of the DNA sequence results of these systems is shown in *Figure 5*. However, the sequence analysis in these two cases has not led immediately to biochemical or molecular understanding, so these are presented here as examples of the limitations of DNA analysis.

The tellurium resistance determinant consists of five open reading frames (ORFs) which appear to govern five proteins, running from 13−38 kd in molecular mass. Whereas the first two of these gene products appear to be involved in regulation (Jobling and Ritchie, 1988), how this might work is as yet unknown. The last three gene products may be involved in the resistance mechanism itself (Jobling and Ritchie, 1988).

For the copper resistance determinant (Mellano and Cooksey, 1988), the DNA sequence of a 4.5 kb segment contains four ORFs (*Figure 5*), identified as probable

Regulation                                Tellurium Resistance

ORF1      ORF2      ORF3      ORF4      ORF5
37.2kDa    13.7kDa   38.2kDa   20.4kDa   20.4kDa

Copper Resistance

ORFA          ORFB      ORFC      ORFD
67.2kDa       36.3kDa   13.0kDa   33.0kDa

**Figure 5.** Diagram of the open reading frame from the DNA sequence analysis of the tellurium resistance determinant (Jobling and Ritchie, 1988) (top) and of the copper resistance determinant (Mellano and Cooksey, 1988) (bottom).

genes. Deletions of or mutations in ORFA and ORFB led to copper sensitivity. ORFC and ORFD are required for full, but not for partial, resistance (Mellano and Cooksey, 1988). A tandemly repeated octapeptide [Asp−His−Ser−Gln(or Lys)−Met−Gln−Gly−Met] occurs five times towards the beginning of ORFB and related octapeptides occur four times in the middle region of ORFA. The methionine sulphurs and histidine imidazole nitrogens were considered candidates for copper-binding residues (Mellano and Cooksey, 1988). The first three gene products appear (from the sequences) to be soluble polypeptides, although they contain hydrophobic N-terminal segments characteristic of membrane signal sequences. Sequence homology to known regulatory proteins led to the hypothesis that the ORFC product may be a positively acting regulatory protein (Mellano and Cooksey, 1988). The ORFD product has several potential membrane-spanning hydrophobic stretches, making the ORFD product a candidate for a copper transport protein (Mellano and Cooksey, 1988). These hypothesized roles for the four copper-resistant ORFs from sequence analysis are less convincing than those described above for the cadmium, arsenic and mercury resistance systems.

With the cloning and analysis of a second copper resistance determinant (Rouch *et al.*, 1989), the potential for better understanding of copper resistance mechanisms from DNA sequencing grows. The four genes of this second copper resistance determinant do not, however, show similarity in apparent size and location (Rouch *et al.*, 1989) to those of the first system. It appears in this case as if the second gene is the regulatory gene and the fourth (smaller) gene is involved in copper binding (Rouch *et al.*, 1989). Once the sequence analysis of this second copper resistance determinant is reported, then we will be able to judge whether two quite separate systems are being studied or not.

In summary, the DNA sequence analysis of the first plasmid resistance systems for cadmium and for arsenic and of six distinct genetic determinants of mercury resistance has provided sophisticated understanding of the physiological and biochemical mechanisms of these resistances. This is very similar to the impact of DNA sequence analysis in other areas of molecular biology. For many additional heavy metal resistance determinants, DNA sequence analysis has yet to be effective. Two more sequencing projects are currently under way in our laboratory; one for the determinant of cadmium, zinc and cobalt resistance in *Alcaligenes* (Nies *et al.*, 1987; Nies and Silver, 1989) and the other for the chromate resistance system in *Pseudomonas* (Ohtake *et al.*, 1987). The application of such modern tools as recombinant DNA analysis will continue to radically advance our understanding of bio-inorganic chemistry.

## Acknowledgements

The research in this report has been supported in part by grants from the National Institutes of Health and the National Science Foundation.

## References

Begley,T.P., Walts,A.E. and Walsh,C.T. (1986a) Bacterial organomercurial lyase: overproduction, isolation and characterization. *Biochemistry* **25**, 7186−7192.

Begley,T.P., Walts,A.E. and Walsh,C.T. (1986b) Mechanistic studies of a protonolytic organomercurial cleaving enzyme: bacterial organomercurial lyase. *Biochemistry* **25**, 7192−7200.

Brandl,C.J., Green,N.M., Korczak,B. and MacLennan,D.H. (1986) Two $Ca^{2+}$ ATPase genes: homologies and mechanistic implications of deduced amino acid sequences. *Cell* **44**, 597−607.

Brown,N.L., Misra,T.K., Winnie,J.N., Schmidt,A., Seiff,M. and Silver,S. (1986) The nucleotide sequence of the mercuric resistance operons of plasmid R100 and transposon Tn*501*: further evidence for *mer* genes which enhance the activity of the mercuric ion detoxification system. *Molecular and General Genetics* **202**, 143 – 151.

Chen,C.M., Misra,T.K., Silver,S. and Rosen,B.P. (1986) Nucleotide sequence of the structural genes for an anion pump. The plasmid encoded arsenical resistance operon. *Journal of Biological Chemistry* **261**, 15030 – 15038.

Distefano,M.D., Au,K.G. and Walsh,C.T. (1989) Mutagenesis of the redox-active disulfide in mercuric ion reductase: catalysis by mutant enzymes restricted to flavin redox chemistry. *Biochemistry* (in press).

Feng,D.-F. and Doolittle,R.F. (1987) Progressive sequence alignment as a prerequisite to correct phylogenetic trees. *Journal of Molecular Evolution* **25**, 351 – 360.

Foster,T.J. (1987) The genetics and biochemistry of mercury resistance. *CRC Critical Reviews in Microbiology* **15**, 117 – 140.

Foster,T.J. and Brown,N.L. (1985) Identification of the *merR* gene of R100 by using *mer-lac* gene and operon fusions. *Journal of Bacteriology* **163**, 1153 – 1157.

Griffin,H.G., Foster,T.J., Silver,S. and Misra,T.K. (1987) Cloning and DNA sequence of the mercuric- and organomercurial-resistance determinants of plasmid pDU1358. *Proceedings of the National Academy of Sciences of the United States of America* **84**, 3112 – 3116.

Jackson,W.J. and Summers,A.O. (1982) Biochemical characterization of HgCl$_2$-inducible polypeptides encoded by the *mer* operon of plasmid R100. *Journal of Bacteriology* **151**, 962 – 970.

Jobling,M.G. and Ritchie,D.A. (1988) The nucleotide sequence of a plasmid determinant for a resistance to tellurium anions. *Gene* **66**, 245 – 258.

Joerger,R.D. and Bishop,P.E. (1988) Nucleotide sequence and genetic analysis of the *nifB-nifQ* region from *Azotobacter vinelandii*. *Journal of Bacteriology* **170**, 1475 – 1487.

Laddaga,R.A., Chu,L., Misra,T.K. and Silver,S. (1987) Nucleotide sequence and expression of the mercurial-resistance operon from *Staphylococcus aureus* plasmid pI258. *Proceedings of the National Academy of Sciences of the United States of America* **84**, 5106 – 5110.

Lund,P.A. and Brown,N.L. (1987) Role of the *merT* and *merP* gene products of transposon Tn*501* in the induction and expression of resistance to mercuric ions. *Gene* **52**, 207 – 214.

Mellano,M.A. and Cooksey,D.A. (1988) Nucleotide sequence and organization of copper resistance genes from *Pseudomonas syringae* pv. *tomato*. *Journal of Bacteriology* **170**, 2879 – 2883.

Misra,T.K., Brown,N.L., Fritzinger,D.C., Pridmore,R.D., Barnes,W.M., Haberstroh,L. and Silver,S. (1984) Mercuric ion-resistance operons of plasmid R100 and transposon Tn*501*: the beginning of the operon including the regulatory region and the first two structural genes. *Proceedings of the National Academy of Sciences of the United States of America* **81**, 5975 – 5979.

Misra,T.K., Brown,N.L., Haberstroh,L., Schmidt,A., Goddette,D. and Silver,S. (1985) Mercuric reductase structural genes from plasmid R100 and transposon Tn*501*: functional domains of the enzyme. *Gene* **34**, 253 – 262.

Mobley,H.L.T. and Rosen,B.P. (1982) Energetics of plasmid-mediated arsenate resistance in *Escherichia coli*. *Proceedings of the National Academy of Sciences of the United States of America* **79**, 6119 – 6122.

Nakahara,H., Silver,S., Miki,T. and Rownd,R.H. (1979) Hypersensitivity to Hg$^{2+}$ and hyperbinding activity associated with cloned fragments of the mercurial resistance operon of plasmid NR1. *Journal of Bacteriology* **40**, 161 – 166.

Ni'Bhriain,N., Silver,S. and Foster,T.J. (1983) Tn*5* insertion mutations in the mercuric ion resistance genes derived from plasmid R100. *Journal of Bacteriology* **155**, 690 – 703.

Nies,D., Mergeay,M., Friedrich,B. and Schlegel,H.G. (1987) Cloning of plasmid genes encoding resistance to cadmium, zinc and cobalt in *Alcaligenes eutrophus* CH34. *Journal of Bacteriology* **169**, 4865 – 4868.

Nies,D.H. and Silver,S. (1989) Plasmid-determined inducible efflux is responsible for resistance to cadmium, zinc, cobalt and nickel in *Alcaligenes eutrophus*. *Journal of Bacteriology* **171**, (in press).

Novick,R.P., Murphy,E., Gryczan,T.J., Baron,E. and Edelman,I. (1979) Penicillinase plasmids of *Staphylococcus aureus*: restriction-deletion maps. *Plasmid* **2**, 109 – 129.

Nucifora,G., Chu,L., Misra,T.K. and Silver,S. (1989a) Cadmium resistance of *Staphylococcus aureus* plasmid pI258 *cadA* gene results from a cadmium efflux ATPase. *Proceedings of the National Academy of Sciences of the United States of America* (in press).

Nucifora,G., Chu,L., Silver,S. and Misra,T.K. (1989b) Mercury operon regulation by the *merR* gene of the organomercurial resistance system of plasmid pDU1358. *EMBO Journal* (in press).

O'Halloran,T. (1988) Metalloregulatory proteins: metal-responsive molecular switches governing gene expression. In *Metal Ions in Biological Systems* Vol. 25. (ed. H.Sigel), Marcel Dekker, New York (in press).

O'Halloran,T. and Walsh,C. (1987) Metalloregulatory DNA-binding protein encoded by the *merR* gene: isolation and characterization. *Science* **235**, 211–214.

Ohtake,H., Cervantes,C. and Silver,S. (1987) Decreased chromate uptake in *Pseudomonas fluorescens* carrying a chromate resistance plasmid. *Journal of Bacteriology* **169**, 3853–3856.

Pabo,C.P. and Sauer,R.T. (1984) Protein-DNA recognition. *Annual Review of Biochemistry* **53**, 293–321.

Rosen,B.P., Weigel,U., Karkaria,C. and Gangola,P. (1988) Molecular characterization of an anion pump. *Journal of Biological Chemistry* **263**, 3067–3070.

Rouch,D., Lee,B.T.O. and Camakaris,J. (1989) Genetic and molecular basis of copper resistance in *Escherichia coli*. In *Metal Ion Homeostasis: Molecular Biology and Chemistry* (eds. D.Winge and D.Hamer), Alan R.Liss, New York (in press).

Silver,S., Budd,K., Leahy,K.M., Shaw,W.V., Hammond,D., Novick,R.P., Willsky,G.R., Malamy,M.H. and Rosenberg,H. (1981) Inducible plasmid-determined resistance to arsenate, arsenite and antimony (III) in *Escherichia coli* and *Staphylococcus aureus*. *Journal of Bacteriology* **146**, 983–996.

Silver,S. and Keach,D. (1982) Energy-dependent arsenate efflux: the mechanism of plasmid-mediated resistance. *Proceedings of the National Academy of Sciences of the United States of America* **79**, 6114–6118.

Silver,S. and Misra,T.K. (1988) Plasmid-mediated heavy metal resistances. *Annual Review of Microbiology* **42**, 717–743.

Silver,S., Nucifora,G., Chu,L. and Misra,T.K. (1989) Bacterial resistance ATPases: primary pumps for exporting toxic cations and anions. *Trends in Biochemical Sciences* (in press).

Solioz,M., Mathews,S. and Furst,P. (1987) Cloning of the $K^+$-ATPase of *Streptococcus faecalis*. *Journal of Biological Chemistry* **262**, 7358–7362.

Summers,A.O. (1986) Organization, expression and evolution of genes for mercury resistance. *Annual Review of Microbiology* **40**, 607–634.

Thieme,R., Pai,E.F., Schirmer,R.H. and Schulz,G.E. (1981) Three-dimensional structure of glutathione reductase at 2 Å resolution. *Journal of Molecular Biology* **152**, 763–782.

Tynecka,Z., Gos,Z. and Zajac,J. (1981) Energy-dependent efflux of cadmium coded by a plasmid resistance determinant in *Staphylococcus aureus*. *Journal of Bacteriology* **147**, 313–319.

Walderhaug,M.O., Dosch,D.C. and Epstein,W. (1987) Potassium transport in bacteria. In *Ion Transport in Prokaryotes* (eds B.P.Rosen and S.Silver), Academic Press, San Diego: pp. 85–130.

Walsh,C.T., Distefano,M.D., Moore,M.J., Shewchuk,L.M. and Verdine,G.L. (1988) Molecular basis of bacterial resistance to organomercurial and inorganic mercuric salts. *FASEB Journal* **2**, 124–130.

Wang,Y., Moore,M., Levinson,H.S., Silver,S., Walsh,C. and Mahler,I. (1989) Nucleotide sequence of a chromosomal mercury resistance determinant from a broad-spectrum mercury-resistant *Bacillus* sp. *Journal of Bacteriology* **171** (in press).

Weiss,A.A., Silver,S. and Kinscherf,T.G. (1978) Cation transport alteration associated with plasmid-determined resistance to cadmium in *Staphylococcus aureus*. *Antimicrobial Agents and Chemotherapy* **14**, 856–865.

# CHAPTER 5

# Interactions of metal ions with components of bacterial cellwalls and their biomineralization

T.J.BEVERIDGE

*Department of Microbiology, College of Biological Science, University of Guelph, Guelph, Canada N1G 2W1*

## Introduction

Microfossils found in ancient cherts and shales suggest that prokaryotes have inhabited the Earth for a surprisingly long period of time. Cherts from the Warrawoona group in Western Australia resemble present-day stromatolites and indicate that bacteria similar to cyanobacteria existed at least $3.5 \times 10^9$ years ago. Presumably, less advanced forms were present even earlier.

The Earth's biosphere mass is very small when compared to the combined mass of the hydrosphere and the crust; however, when the production rate of organic material ($10^{17}$ g year$^{-1}$) is integrated over time, the total mass recycled by biology approaches the mass of the Earth (Abelson, 1957). Moreover, since average biomass contains metals at the 10 000 p.p.m. level (Trudinger and Swaine, 1979), the total inorganic mass recycled over the same time span approaches the mass of the Earth's crust (Beveridge and Fyfe, 1985). The antiquity of prokaryotes, as compared to eukaryotes, suggests that most biological inorganic recycling through time has been accomplished by bacteria. Clearly, the distribution of bacteria and their ability to recycle inorganic elements must have had (and continues to have) a tremendous effect on the local constitution of sediments, soils and sedimentary rock. With respect to precious metals, it is possible that highly active, yet discriminatory, concentrating systems in ancient bacteria could have contributed to some present-day ore deposits.

This chapter outlines several of the bacterial surfaces we have studied, their electronegative nature, their ability to bind metals, and the eventual mineral which is produced. At the same time, some of the minerals found on bacterial surfaces residing in natural environments are identified, and modern data are used to suggest a feasible route by which ancient prokaryotic microfossils could have been produced.

## Anionic nature of the bacterial surface

Since bacteria must depend entirely on diffusion for their nutrition and removal of their waste products (Beveridge, 1988), there is a prime necessity for their surface to be wettable. This requires an abundance of chemical functional groups to stud the bacterial surface whether it be cell wall, capsule, sheath, or surface array. At the proper growth pH (usually between pH 5 and 8), the groups are ionized and the surface is hydrophilic. Exposed sulphydryl groups are rarely encountered on bacterial surfaces (Beveridge, 1981), and carboxylates, phosphates and amines are the preferred chemical functions for ionization.

Natural waters exist in a variety of salinities, from those saturated with salts, for example, the Dead Sea, to those which contain only trace amounts of salts, for example, rain and dew. However, all contain dissolved metals in some concentration and form. Clearly, since bacteria live in aqueous environments, their charged surfaces must be in contact and in equilibrium with the various ions which abound in their local milieu. For electronegative sites on the surface of bacteria, $H^+$ or $H_3O^+$ are also important cations and can predominate at low pH. Accordingly, their surfaces are continuously in some salt form or another which is dependent on ion concentration, ion speciation and site specificity.

This metal−surface relationship is, of course, a dynamic one, and rapidly alters (according to external diffusion) as the ion content of the environment changes. Motile bacteria can swim from one environment to another. Yet, in an encompassing way, we can envisage that the magnitude of the metal−bacterium interaction is often simply a reflection of the cellular surface area−volume relationship. The surfaces of bacteria can be composed of a variety of structures, such as gram-negative or gram-positive walls, S-layers, capsules, and slime layers; each, in its own way, uniquely contributes to bacteria−metal interactions. In an even more general sense, a rod, such as *Escherichia coli*, has more surface area than a coccus of equal volume, such as *Streptococcus fecium* (Beveridge, 1988) and, therefore, has a larger interface for interaction with external metallic ions. Obviously, the same is true for comma (for example, *Vibrio*), spiral (for example, *Aquaspirillum*), prosthecate (for example *Caulobacter*), ring-shaped (for example, *Anacyclobacter*) and square bacteria (*Figure 1*). More than any other lifeform, because of their small size and high surface area−volume ratio, as distinct cells bacteria are designed to offer the greatest amount of space for interaction with the external milieu (Beveridge, 1988).

*Metallic ions stabilize the outer membrane of Escherichia coli*

The cell envelope of *E.coli* K-12 is representative of the gram-negative eubacterial variety of surface (Beveridge, 1981). It has no capsule or S-layer, and the wall consists of an amalgam of outer membrane and murein residing on top of a plasma (cytoplasmic) membrane. Between the outer and plasma membranes resides the periplasm.

The outer membrane is the outermost structure of the bacterium and exists as a lipid−protein bilayer; its outer face is exposed to the external milieu. Only 2−3 polypeptides reside in the membrane in large numbers, for example OmpC and OmpF as porins, and the more minor polypeptides are frequently dependent on strain or growth. Of importance for this discussion is the fact that there is an asymmetric distribution of lipid across the bilayer. Lipopolysaccharide is found predominantly on the outer membrane face, whereas phospholipid (mostly phosphatidylethanolamine) is distributed over the inner face. Of the two lipids, lipopolysaccharide possesses the greatest number of electronegative sites per molecule and it points towards the external milieu.

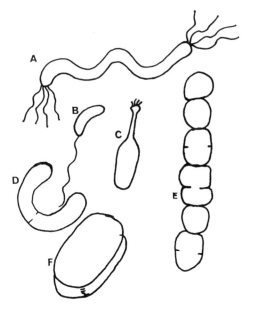

**Figure 1.** Some of the shapes of bacteria are shown; **(a)** spiral (e.g. *Aquaspirillum serpens*), **(b)** comma-shaped (e.g. *Vibrio cholerae*), **(c)** prosthecate (e.g. *Caulobacter crescentus*), **(d)** ring-shaped [e.g. *Anacyclobacter (Microcyclus aquaticus*], **(e)** chains of cocci (e.g. *Streptococcus pyogenes*), and **(f)** square-shaped (e.g. Waslbys' square archaebacterium).

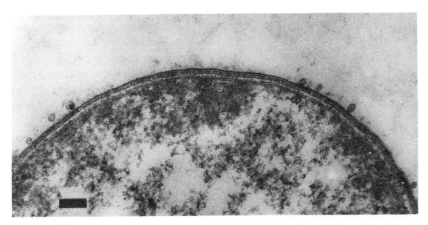

**Figure 2.** Thin section of *E.coli* which has been treated with EDTA according to Ferris and Beveridge (1986a) so that small outer membrane vesicles are sloughing off into the medium. The bar indicates 100 nm.

Over the last decade, studies have shown that the structural layers of the *E.coli* envelope have the capacity to bind metals from solution (Beveridge and Koval, 1981; Hoyle and Beveridge, 1983, 1984). At the same time, metal chelating agents, such as EDTA, have a profound destabilizing effect on the outer membrane in that they extract lipopolysaccharide, phospholipid and protein (Leive, 1965). Electron microscopy has shown that these components are extracted as distinct, small outer membrane vesicles (*Figure 2*). Approximately 33% of the lipopolysaccharide, 24% of the phospholipid

67

and 10% of the protein (as per cent dry weight) can be extracted using the conditions of Ferris and Beveridge (1986a). More importantly, on a dry weight basis, 24% of the constituent outer membrane metal is also removed; 16% of this metal is calcium and 6% is magnesium (Ferris and Beveridge, 1986a). Both of these metallic ions are important for stabilizing the organic charge density within the membrane, particularly the phosphoryl groups of the lipopolysaccharide (Ferris and Beveridge, 1984) since only one carboxylate of the three resident 2-keto-3-deoxyoctonates in the molecule are available for inorganic complexation (Ferris and Beveridge, 1986b).

The EDTA release of outer membranous material is a striking example of the importance of metallic ions for the stability of bacterial surface structures. In this particular case, the outer lipid face of the bilayer is predominantly lipopolysaccharide which possesses approximately 75% of the total phosphate in the outer membrane. These phosphoryl groups are highly electronegative at growth pH, point towards the external environment, and interact strongly with available $Ca^{2+}$ and $Mg^{2+}$. Once complexed to the molecule, these divalent cations become an essential component part of the bilayer structure where they affect molecular motion and produce an optimal lipid packing order for this natural membrane. Clearly, the same holds true for the phospholipids in the membrane, but they possess only one phosphoryl group per molecule. We also suspect that charge neutralization and salt-bridging are important for outer membrane protein conformation and the adhesion of boundary lipid to these proteins. Indeed, the organic building blocks of the outer membrane are so designed that magnesium and calcium are preferred as an inorganic cement and, in fact, membrane stability is a function of the interdependence between the organic and inorganic constituents on each other. When EDTA extracts calcium and magnesium, the motion and packing order of the lipids are so disturbed that they require a more highly curved surface than that provided by the bacterium; in a thermodynamic sense, they have no option but to form small (highly curved) vesicles which can no longer reside in the outer membrane (*Figure 2*). Since lipopolysaccharide is more highly charged than phospholipid, it is more strongly affected (33% lipopolysaccharide versus 24% dry weight phospholipid are extracted). Protein is so dependent on the lipid constituents that it, too, is extracted, but to a lesser extent (10% dry weight).

The idea that metallic ions are an essential part of the outer membrane of gram-negative eubacteria has many implications. The strength of the bonding between metal ion and surface ligand is primarily electrostatic and will depend on a number of factors such as field strength, ion speciation, ionic radii and surface accessibility. It is a dynamic relationship and constituent ions can be displaced and replaced by others from the external milieu; these need not be only other metallic ions. For example, electropositive aminoglycoside antibiotics, such as gentamicin and amikacin, replace essential calcium and magnesium in the outer membrane of *Pseudomonas aeruginosa* thereby disturbing membrane stability and extracting small vesicles. Consquently, small, transient holes are produced which allow the antibiotics easier access to the plasma membrane where they are transported into the cell to disrupt protein synthesis (Bryan and van den Elzen, 1985; Martin and Beveridge, 1986; Walker and Beveridge, 1988).

Even though it is apparent that outer membrane integrity is dependent on its constituent metals, the ability of metallic ions to neutralize surface chemical groups points to another important property of bacterial surfaces, the ability of the surface to limit access of

**Table 1.** Partition coefficients of *Escherichia coli* K-12 cells exposed to EDTA and/or aqueous metal salt solutions before partitioning in a biphasic system of dextran and polyethyleneglycol.

| Metal | Partition coefficient (K)[a,b] | |
|---|---|---|
| | Control | EDTA |
| — | 0.20 | 0.37 |
| Na | 0.18 | 0.24 |
| Mg | 1.13 | 0.88 |
| Ca | 0.68 | 0.69 |
| Mn | 1.43 | 0.75 |

[a]Calculated as defined in Ferris and Beveridge (1986a) from which the table was taken with permission of the authors and the National Research Council of Canada.
[b]Figures represent averages of triplicate samples derived from three separate 1 l cultures.

the external aqueous milieu. In a general sense, since most bacterial surfaces are charged (Beveridge, 1981), they are hydrophilic and, therefore, wettable. For organisms which depend on the diffusion of nutrients and wastes for their very existence, this surface property is not merely a convenience, it is of vital importance. Yet, since dissolved salts abound in their environment, surface charge neutralization is also a fact of life. Biphasic partition experiments have shown that select metallic ions can either increase or reduce the hydrophobicity of *E. coli* (*Table 1*). Control cells are highly hydrophilic (wettable, $K = 0.20$), whereas Mn-cells are hydrophobic (unwettable, $K = 1.43$). Pretreatment with EDTA, once again, points to the importance of metal−lipopolysaccharide interaction for the phenomenon (*Table 1*).

Obviously, there are select regions of the *E. coli* outer membrane which will be continuously open to diffusion (for example, the water-filled porin channels), but the salt form of the outer membrane must affect surface wettability. If it is possible that the bacterium has some control over this property, many advantages are apparent. For example, a high degree of hydrophobicity can be used to help the bacterium contact unwettable surfaces and for their adherence to them. This may aid in microcolony formation on inert surfaces, in the utilization of apolar hydrocarbons for nutrition, and in the exclusion and prevention of solvated particles such as bacteriophages from adhering to the bacterium. Hydrophilicity, on the other hand, gives the bacterium freer access to the surrounding water and its solutes, and allows easier exchange. The possibilties are wide ranging and affect such diverse topics in microbiology as ecology, medicine and biotechnology.

## The charge distribution of the sheath of Methanospirillum hungatei

Unlike eubacterial surfaces, few studies have been done on the charge distribution of archaebacterial surfaces. *Methanospirillum hungatei* is a methanogen which metabolically converts hydrogen and carbon dioxide to methane. Its structural organization differs considerably from that of most other bacteria (Zeikus and Bowen, 1975) since its external sheath, instead of its wall, provides cellular shape and resists turgor pressure (Beveridge *et al.*, 1987; Sprott *et al.*, 1979). The proteinaceous sheath has been resolved to extremely high resolution (Stewart *et al.*, 1985) and a three-dimensional projection made (Shaw *et al.*, 1985). For this reason, the sheath has been a suitable structure to map

**Figure 3.** Negative stain of *M.hungatei* sheath which has been treated with polycationic ferritin for the surface labelling of anionic sites on the sheath. The polycationic ferritin is the small, doughnut-shaped particle. The bar indicates 100 nm.

charge density to a resolution approaching its crystalline order.

Polycationic ferritin (PCF) was first used as a macromolecular probe to determine if the sheath was anionic (*Figure 3*). Thin sections revealed that both inner and outer surfaces of the sheath were studded with PCF. Chemical neutralization of carboxylate groups after carbodiimide activation with glycine methyl ester inhibited PCF adsorption and suggested that $COO^-$ dominated the electronegative character of the surfaces M.Sára, U.Sleytr and T.J.Beveridge, unpublished observations).

The large size of PCF (10 nm in diameter) made high-order resolution of the charge distribution on the sheath impossible and a smaller probe was necessary. Cytochrome *c* is a small (12.4 kd), ellipsoidal (3.7 × 2.5 × 2.5 nm) protein which, because of its high isoelectric point (pI = 10.8), is electropositive at neutral pH (Takano and Dickerson, 1981). When this small molecule was used to probe the electronegative sites and the sheath negatively stained for electron microscopy there was little apparent change between the control and cytochrome *c* images. Yet, optical transforms of the cytochrome *c* images revealed a wealth of detail not present in control preparations, and computer reconstruction of the image showed that the probe resided in the spaces between the proteinaceous particles on the sheath surface (*Figure 4*) (D.Pum, M.Sára, U.Sleytr and T.J.Beveridge, unpublished observations). It is very possible that these are the regions which are rich in $COO^-$.

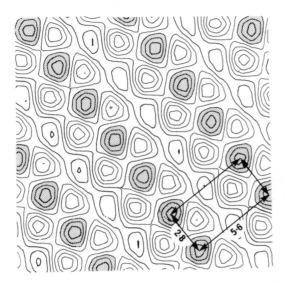

**Figure 4.** Contour plot of a computer enhanced image of cytochrome *c* adhering to the *M.hungatei* sheath. The cytochrome *c* has a darker colour than the proteinaceous particles of the sheath, sits between the sheath particles, and has a 2.8 × 5.6 nm unit cell. (Courtesy of D.Pum, Zentrum für Ultrastrukturforschung der Universität für Bodenkultur, Wein.)

From a mechanistic view, this location for electronegative sites on the sheath is convenient. The sheath, because of its minute particle spacing, should be a relatively impervious structure which allows only small molecules to pass through (for example, molecular hydrogen, carbon dioxide and methane); the channels through the sheath would occur between its resident particles (Stewart *et al.*, 1985).

Although hydrogen, carbon dioxide and methane are electroneutral (van der Waals forces would allow small, transient dipole moments), their solubility is enough to allow growth of the bacterium; they must penetrate through the sheath. At the same time, *M.hungatei* has a need for metallic ions which also must pass through. Clearly, the cell possesses a Na/K gradient across its plasma membrane, and magnesium and calcium are required for membrane and ribosome stability. Furthermore, at least three metalloenzymes (a hydrogenase, a carbon monoxide dehydrogenase and a methyl coenzyme M reductase) require nickel for their activity (Sprott *et al.*, 1987). Recent experiments have shown that the sheath of *M.hungatei* and that of a related methanogen, *Methanothrix concilii*, are proficient at binding nickel under reducing conditions (*Figure 5*) (G.Southam, G.P.Patel, G.D.Sprott and T.J.Beveridge, unpublished observations). These results suggest that the sheath concentrates nickel from solution and may even act as a reservoir for the metal from which the cell can draw.

*Interaction of metallic ions with Bacillus subtilis walls*

The walls of *Bacillus subtilis* 168 are capable of binding more metal than any other system we have studied. When they are grown in medium possessing relatively high concentrations of phosphate, their walls are virtually a two-polymer system of peptidoglycan (approximately 46% dry weight) and teichoic acid (approximately 54%

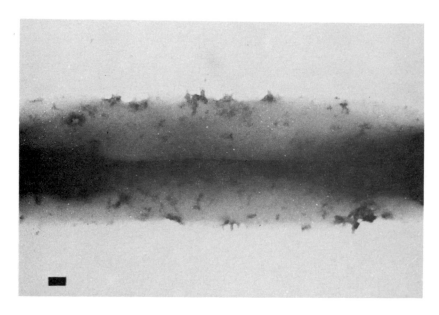

**Figure 5.** Whole mount of *Methanothrix concilii* which has been pretreated with nickel in a Tris−HCl/H₂S buffer. No stain has been used and the electron density is derived from the sorbed nickel. The bar indicates 100 nm. (Courtesy of G.Southam, University of Guelph, and G.Patel and G.D.Sprott, NRC of Canada.)

dry weight). If the metal salts in the medium are kept to a minimum (for example, the ionic balance maintained primarily with $NH_4^+$ and $Cl^-$ with just enough $Na^+$, $K^+$, $Mg^+$ and trace heavy metals to sustain growth) the walls are so depleted of metals that their electron-scattering profile is less than the embedding plastic which surrounds them in thin section (*Figure 6*). Yet, using PCF as an electropositive probe for electron microscopy we obtain substantial binding to the wall since it is so anionic (Sonnenfeld *et al.*, 1985a). Yet, the pattern of binding is strikingly different from that of most bacilli; PCF adheres only to the outer surface of the wall (*Figure 7*) which indicates that there is an asymmetrical charge distribution between the wall's inner and outer surfaces (Sonnenfeld *et al.*, 1985a). Furthermore, when dilute, limiting concentrations of PCF are used, only regions at the polar ends of the walls are labelled, suggesting that these areas are more electronegative than others (Sonnenfeld *et al.*, 1985b).

This wall, as a two-polymer system, has only a set number of chemically active sites to which metallic ions can bind over the range of pH values at which the bacterium can grow. For teichoic acid, the phosphate groups will be preferred; for peptidoglycan, the carboxylates of the peptide stems will be preferred (*Figure 8*). It is a good model system to work with since the amount of each polymer is easily established and the exact number of chemically reactive groups can be quantified; work from two laboratories clearly established that the peptidoglycan rather than the teichoic acid was responsible for most of the metal deposition (Beveridge and Murray, 1980; Doyle *et al.*, 1980). Certainly, when $COO^-$ groups within the wall are chemically neutralized or reversed, the amount of metal which is deposited is substantially reduced (*Table 2* is representative).

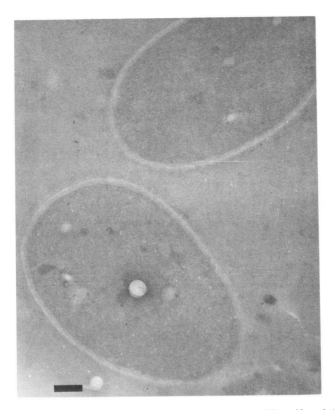

**Figure 6.** Thin section of two *B.subtilis* cells grown under metal limiting conditions. No stain has been used and the cell walls have less electron scattering power than the surrounding plastic or the cytoplasm. The bar indicates 100 nm.

A clear distinction is made between 'deposited' and 'bound' in order to emphasize the complexity of metal sorption to these walls. Far more metal is deposited than the expected stoichiometry of metal to reactive wall site would have us believe. Electron microscopy of the walls showed that, frequently, large metallic deposits studded their structure (*Figure 9* is an extreme example). Our explanation for this is that it is at least a two-step deposition process. The first step in time would be the stoichiometric binding of metal to reactive group within the wall. This, then, would nucleate the deposition of more metal, as a mineral, at the same location (Beveridge and Murray, 1976). The size of the mineral would depend on a number of variables, the most obvious of which are the concentration of the metallic ion in solution and the amount of time through which the reactions proceed. At the same time, only a limited amount of space is available between polymers within the wall matrix (Beveridge, 1981) to accommodate mineral growth unless covalent bonds are broken. Growth at the wall surfaces does not have this restriction and deposits can grow to a large size. In fact, once the solute metal is depleted, it is not unusual to find that surface minerals grow at the expense of those deposits within the wall fabric; the one gets larger while the other gets smaller. Depending on the metal ion supplied to the wall, the mineral can be entirely metallic (for example, gold as in *Figure 9*), or as a complex hydrated oxide (Beveridge and Murray, 1976).

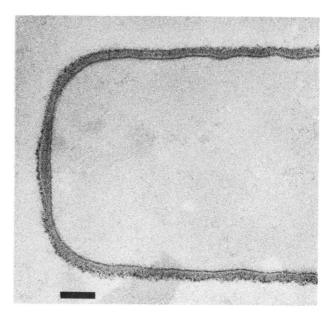

**Figure 7.** Thin section of a *B.subtilis* cell wall treated with polycationized ferritin (small electron dense spots). The polycationized ferritin has stuck only to the outer surface of the wall, suggesting an asymmetric distribution of anionic sites. The bar indicates 100 nm.

## Observations from natural sediments

It is important, now that we have determined that representatives of gram-negative and gram-positive eubacterial walls and of archaebacterial walls can bind substantial quantities of metal, that we ascertain if there are important consequences for these observations in nature. To begin these studies, we travelled to the Norris Geyser Basin in Yellowstone National Park, Wyoming, USA to sample sediments from select geothermal settings. This offered the advantage of being able to choose hot springs of distinct pH and inorganic solute load, as well as ensuring that the extreme temperatures would enrich for bacteria and restrict the numbers of eukaryotic microorganisms in the sediments.

Cyanidium Creek is an acidic stream (pH 3.1) which contains high levels of dissolved metals and produces surficial deposits of clay and siliceous mud. A significant portion of the organic matter in the sediment is microbial, and is derived from the eukaryotic alga *Cyanidium caldarium* (the only thermophilic photoautotroph of acid springs) and from thermoacidophilic bacteria (Brock, 1978). Examination of the sediments (temperature 60°C) by electron microscopy clearly showed advanced mineralization on bacterial surfaces (*Figure 10*). Energy dispersive X-ray spectroscopy (EDS) of these minerals demonstrated that they were primarily an iron silicate (Fe 19.7% and Si 55.5% by weight) with varying levels of arsenic, sulphur, aluminium and potassium associated with them (Ferris *et al.*, 1986). In general, the greater the depth of the sampled sediment, the greater the degree of mineralization; some samples showed only the mineralized remains of sacculi and little organic structure was visible. We believe that in this oxic

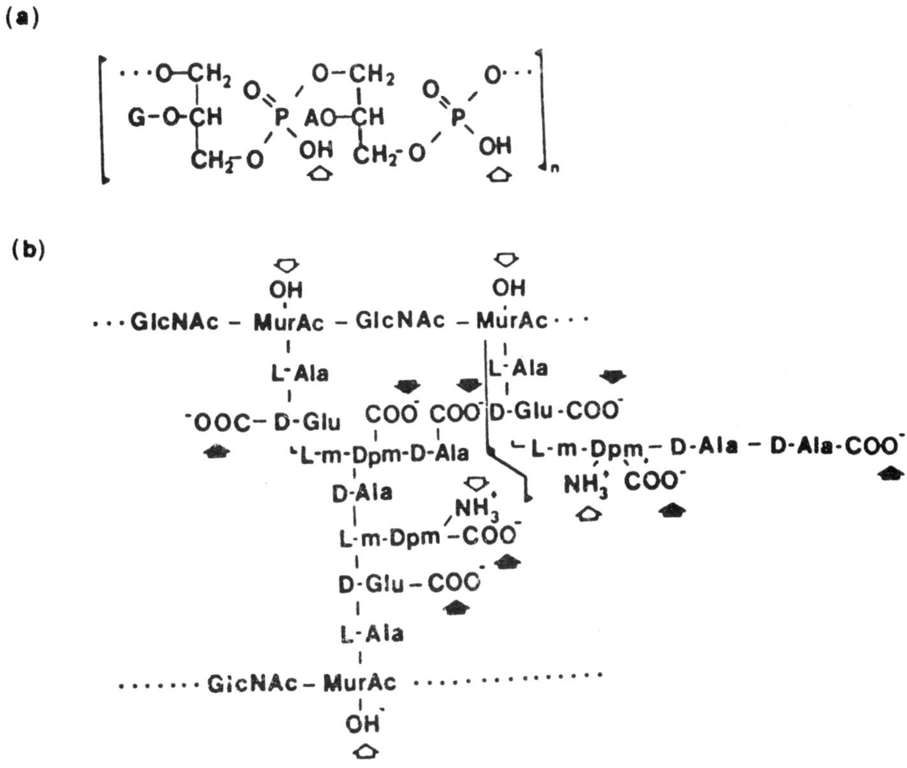

**(a)**

**(b)**

**Figure 8.** (a) Structure of the teichoic acid in the *B.subtilis* wall which may be partially substituted with ester-linked D-alanine (D) and which contains D-glucose (G) attached to the glycerol backbone. Arrows point to the ionizable groups of the phosphodiester bonds $n = \sim 20-30$ residues. (b) Structure of the peptidoglycan which shows both the transpeptide-linked and unlinked conditions. About 35% of the glycan strands are cross-linked, and, in this case, five carboxyl and one amine group should be available. Each unlinked pentapeptide has three carboxyl and one amine group. Some ionizable groups are indicated; the solid arrows indicate the carboxyl and the open arrows point to other available groups. All abbreviations are as usual except L-m-Dpm=(L)-meso-diaminopimelic acid. Reprinted by permission of the authors and the American Society for Microbiology.

environment, microdomains of iron oxides, carbonates and silicates would develop from immobilized hydroxide precursors associated with the bacterial surface. As diagenesis progresses, the cells would die and lyse leaving the mineralized surface to develop according to the geochemical environment over time.

At another location in the Norris Geyser Basin we sampled a hot spring, Terrace Spring, which had high levels of dissolved manganese. In these sediments (temperature 48°C, pH 7.1) the dissolved manganese concentration decreased with sample depth (Ferris *et al.*, 1987a). This was surprising since the reducing conditions of the deep sediment levels should remobilize the manganese oxides formed in the upper oxic region; something was altering the expected geochemistry of manganese as it progressed into the sediment. Electron microscopic and EDS examination revealed that bacterial surfaces were becoming encrusted with manganese oxides as the sampling depth increased, and

**Table 2.** Effect of reversing the charge of the carboxylates in *Bacillus subtilis* walls on metal uptake[a].

| Metal (as chloride salts) | Control walls[b] ($\mu$mol metal per mg wall) | Ethylenediamine walls[c] ($\mu$mol metal per mg wall) |
|---|---|---|
| Na | 2.697 | 0 |
| K | 1.944 | 0 |
| Mg | 8.226 | 0.160 |
| Ca | 0.399 | 0.300 |
| Mn | 0.801 | 0.100 |
| Fe(III) | 3.581 | 0.240 |
| Ni | 0.107 | 0.004 |
| Cu | 2.990 | 0.260 |
| Au | 0.363 | 0.018 |

[a]For details of the experiment refer to Beveridge and Murray (1980).
[b]Control walls were tested for their ability to immobilize metal from solution according to Beveridge and Murray (1976).
[c]These walls were treated exactly as the control walls except they had their $COO^-$ groups reversed to $NH_3^+$ by the linkage of ethylenediamine (Beveridge and Murray, 1980).

**Figure 9.** Thin section of a *B.subtilis* wall which has been treated with $AuCl_3$ according to Beveridge and Murray (1976); it has become totally encrusted with elemental gold. The bar indicates 100 nm.

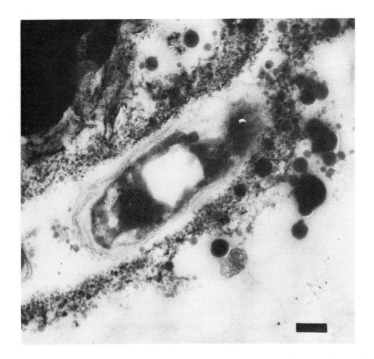

**Figure 10.** Thin section of a bacterium from Cyanidium Creek which is surrounded by an iron−silica mineral. The bar indicates 100 nm. (Reprinted with permission of the authors and the publisher.)

that an unusual manganese oxide polymorph was forming. This was todorokite, which has an O/Mn ratio of 1.7:1.9, a 0.244 nm d-spacing, and a prominent 1.0 nm tunnel structure (*Figure 11*) (Turner and Buseck, 1981; Ferris *et al.*, 1987a). In this manganate, chains of three edge-shared $MnO_6$ octahedra form the sides of the tunnels which, when viewed perpendicular to the tunnel direction, form a fibrous crystal lattice which is often twinned (overlapped lattices) at 120°; the tunnels are usually filled with water containing dissolved calcium, sodium or potassium (Piper *et al.*, 1984). As with the iron−silica situation of Cyanidium Creek sediments, here we have another sediment whose geochemistry is altered, and possibly even dominated, by bacterial surfaces.

Since the Yellowstone samplings, several other natural settings have been examined. Space does not permit a description of the results from all locations and only two will be briefly described.

The first is important since it is from an undisturbed shale/mudstone dating back 39 000 years (*Figure 12*). The sample is from the Cold Regions Research and Engineering Laboratory tunnel, near Fairbanks, Alaska, which was originally dug to assess the methods and effects of mining through permafrost (Sellman, 1972). It has been maintained in a frozen state since it was dug during the Second World War. Most of the tunnel is through a former river bed and extends into a loess from the Yukon−Tanana Uplands which was reworked by the river into a deposit of silt and gravel called the Wisconsin Silt. Electron microscopy of this ancient sediment demonstrated a range of large clastic particles (greater than 10 $\mu$m in diameter)

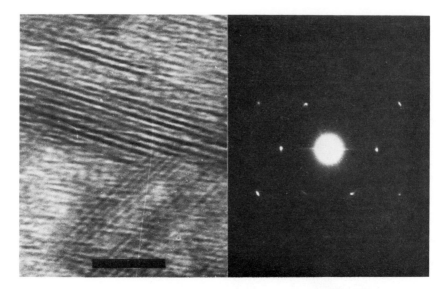

**Figure 11.** (a) High-resolution electron micrograph of a thin manganese oxide crystal showing the prominent 1 Å tunnel structure and twinning of overlapping fibres at 120° angles (the bar represents 10 nm). (b) Selected area electron diffraction pattern from the manganese oxide crystal shown in (a). The first strong reflections have a 2.44 Å *d*-spacing. (Reprinted with the permission of the authors and the publisher.)

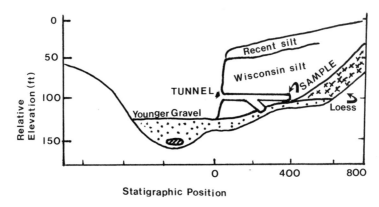

**Figure 12.** Sketch of the Cold Regions Research and Engineering Laboratory, Fairbanks, Alaska. Site of sample collection is indicated.

intermixed with a fine clay-like dispersion. Close examination of the dispersion revealed it to be interspersed with bacteria and their surface remnants (*Figure 13*); once again, authigenic mineral development was proceeding. It is therefore apparent that mineral formation by bacterial surfaces is not just a recent affair; it has been occurring over geologically significant time frames.

The last natural setting to be described suffers from anthropogenic heavy metal loading. In the Onaping region near Sudbury, Ontario, Canada, there are large tailing ponds derived from the local nickel mines. These seep into Upper Moose Lake and provide a substantial heavy metal load. Oxidation of sulphur lowers the pH of the lake to 3.0.

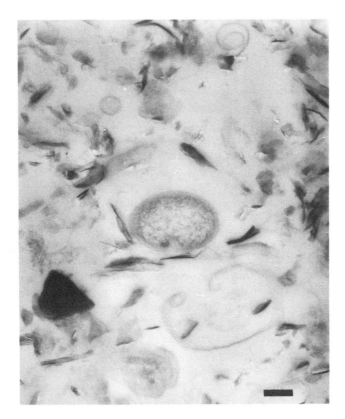

**Figure 13.** Thin section of the sample taken from the Cold Regions Research and Engineering Laboratory. Note the fine, clay-like material which surrounds and is associated with the bacteria and their products. The bar indicates 100 nm.

This water is treated with crushed limestone before it is allowed to flow into a sister lake (Lower Moose Lake) whose pH remains at a constant 7.0. It was the lower lake that we sampled.

In this type of situation, we would expect that a bacterially-driven dissimilatory reduction of sulphate to sulphide would occur in the anoxic sediments of Lower Moose Lake. Because soluble iron and nickel are high in the waters, this sulphide should complex to form Fe/Ni sulphides as the major mineral. Although millerite (NiS) and mackinawite ($FeS_{1-x}$) were found associated with bacteria in the sediments, these minerals were infrequent (Ferris *et al.*, 1987b). Clays (iron and aluminium silicates) were more abundant and ranged from granular (limonitic) to interstratified crystalline (chamositic) varieties with various intermediate stages between the two. It was apparent that there was not enough organic carbon (0.75% dry weight) in the sediments to maintain sulphate reduction and sustain abundant metal sulphide production, and that complex Fe, $AlSiO_n$ polymorphs were preferred (Ferris *et al.*, 1987b). Presumably, soluble iron and aluminium partition themselves from the fluid phase on to the bacterial surfaces where they provide suitable interfaces for the precipitation of silica. The early stages of clay production would therefore appear as fine, granular, limonite phases which, through time, would become more and more crystalline until interstratified chamosite [$(Fe_5Al)(Si_3Al)O_{10}OH_8$] developed.

It is therefore becoming apparent that, in nature, bacteria contribute in a substantial way to the initiation and development of minerals within sediments. Not only do they speed up authigenesis by providing a suitable interface for the capture of metallic ions, they also seem to direct sorption towards specific mineral phases which can be quite unique (for example, todorokite). At the same time, the growth of very complex minerals, such as clays, can be instigated and brought to relative maturity (for example, chamosite). These are startling discoveries and may help to explain the relative abundance of some minerals in the Earth's crust.

## Implications for the fossilization of microorganisms

Microfossils have been identified in ancient sedimentary horizons which date back as far as $3.5 \times 10^9$ years ago (for example, the stromatolitic chert of the Warrawoona group in Western Australia). Since bacteria are made entirely of 'soft tissue' which should not stand the test of time, it is important that we understand how the ancient prokaryotic lifeforms could have been preserved in the rock record. Indeed, this would help verify that these so-called microfossils are, in fact, remnants of ancient life.

Clearly, the work we have been discussing in this chapter is relevant to understanding the fossilization of bacteria. There is reason to believe that if the mineralization of present-day bacteria is an ongoing process, it could have been happening in ancient times too. Since microfossils are found in cherts (a common siliceous sedimentary rock which is a form of quartz) and shales (a consolidated form of mud and clay), it is necessary to account for the silicification of the cells prior to lithification. Several reports have suggested that the prior incorporation of metals into microbial cells favours the preservation of their intact structure (Golubic and Barghoorn, 1977; Southgate, 1986), but the exact reasons for this are not clear.

Using *B.subtilis* cells as a model towards silicic fossilization, we have come close to understanding the series of early mineralizations necessary for the accurate preservation of bacteria. Earlier experiments with *B.subtilis*, which simulated a low-temperature (100°C) diagenesis of organic-rich sediments, used metal-loaded cells in a synthetic sediment of quartz or calcite, and established that the bacteria tenaciously retained the metals (primarily in their walls) during early diagenesis (Beveridge *et al.*, 1983). These experiments also showed that some semblance of form was retained as the bacteria became mineralized. Yet, the preservation was not as good as we would have liked and silification was not possible under the conditions which were used.

More recently, we have determined that there are two prime requirements for the accurate preservation of these bacteria (Ferris *et al.*, 1987c). First, heavy metal pretreatment (for example with $Fe^{3+}$) of the cells saturates the chemically reactive sites in the wall and makes them resistant to degrading autolysins. This is an important step since it is the wall which maintains the shape of the cell and it is the most resilient structure within the bacterium (Beveridge, 1981); without heavy metal pretreatment the shape of the cell frequently degrades.

The second step is the silicification of the cell and is necessary for the eventual entire mineralization of the bacterium. Since $Fe^{3+}$ pretreatment has preserved the wall (and

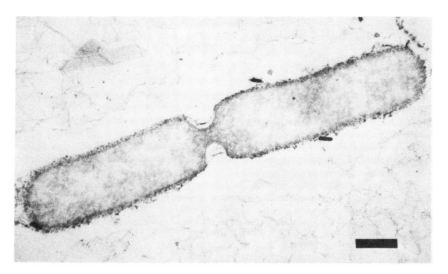

**Figure 14.** Thin section of an iron-loaded, dividing cell of *B.subtilis* which has been aged in the presence of silica for 90 days. The wall has been mineralized completely by silica. The bar indicates 500 nm. (Reprinted with permission of the authors and the publisher.)

bacterial shape), this surface is available for the growth and development of silica crystallites (Ferris *et al.*, 1987c). Through time these grow until the entire surface is encrusted and accurate shape is ensured (*Figure 14*). Silicification can now spread throughout the entire fabric of the cell, and the end result is a microfossil. The entire process is part of the lithification of the sediment and, depending on its geochemical environment through time as a sedimentary rock, a more mature rock such as shale or chert results.

## Acknowledgements

For several aspects of this work, I am indebted to F.G.Ferris for his hard work and enthusiasm while a graduate student and a postdoctoral fellow. The work described from the author's laboratory was supported by operating grants from the Natural Science and Engineering Council (NSERC) of Canada and the Medical Research Council (MRC) of Canada. The *Methanospirillum* research on surface charge was done at the Zentrum für Ultrastrukturforschung der Universität für Bodenkultur, Wien, Austria with the able help of U.Sleytr, P.Messner, D.Pum and M.Sára while T.J.B. was on sabbatical leave with the aid of a NSERC/Austrian Science Foundation Bilateral stipend and a MRC travel grant.

## References

Abelson,P. (1957) Some aspects of paleobiochemistry. *Annals of the New York Academy of Sciences* **69**, 276–285.
Beveridge,T.J. (1981) Ultrastructure, chemistry, and function of the bacterial wall. *International Review of Cytology* **72**, 229–317.
Beveridge,T.J. (1988) The bacterial surface: general considerations towards design and function. *Canadian Journal of Microbiology* **34**, 363–372.

Beveridge,T.J. and Murray,R.G.E. (1976) Uptake and retention of metals by cell walls of *Bacillus subtilis*. *Journal of Bacteriology* **127**, 1502−1518.

Beveridge,T.J. and Murray,R.G.E. (1980) Sites of metal deposition in the cell wall of *Bacillus subtilis*. *Journal of Bacteriology* **141**, 876−887.

Beveridge,T.J. and Koval,S.F. (1981) Binding of metals to cell envelopes of *Escherichia coli* K-12. *Applied and Environmental Microbiology* **42**, 325−335.

Beveridge,T.J. and Fyfe,W.S. (1985) Metal fixation by bacterial cell walls. *Canadian Journal of Earth Science* **22**, 1892−1898.

Beveridge,T.J., Harris,B.J. and Sprott,G.D. (1987) Septation and filament splitting in *Methanospirillum hungatei*. *Canadian Journal of Microbiology* **33**, 725−732.

Beveridge,T.J., Meloche,D.J., Fyfe,W.S. and Murray,R.G.E. (1983) Diagenesis of metals chemically complexed to bacteria: laboratory formation of metal phosphates, sulfides, and organic condensates in artificial sediments. *Applied and Environmental Microbiology* **45**, 1094−1108.

Brock,T.D. (1978) *Thermophilic Microorganisms and Life at High Temperatures*. Springer-Verlag, New York.

Bryan,L.E. and van den Elzen,H.M. (1975) Gentamicin accumulation by sensitive strains of *Escherichia coli* and *Pseudomonas aeruginosa*. *Journal of Antibiotics* **28**, 696−703.

Doyle,R.J., Matthews,T.H. and Streips,U.N. (1980) Chemical basis for selectivity of metal ions by the *Bacillus subtilis* cell wall. *Journal of Bacteriology* **143**, 471−480.

Ferris,F.G. and Beveridge,T.J. (1984) Binding of a paramagnetic metal cation to *Escherichia coli* K-12 outer membrane vesicles. *FEMS Microbiology Letters* **24**, 43−46.

Ferris,F.G. and Beveridge,T.J. (1986a) Physicochemical roles of soluble metal cations in the outer membrane of *Escherichia coli*. *Canadian Journal of Microbiology* **32**, 594−601.

Ferris,F.G. and Beveridge,T.J. (1986b) Site specificity of metallic ion binding in *Escherichia coli* K-12 lipopolysaccharide. *Canadian Journal of Microbiology* **32**, 52−55.

Ferris,F.G., Beveridge,T.J. and Fyfe,W.S. (1986) Iron-silica crystallite nucleation by bacteria in a geothermal sediment. *Nature (London)* **320**, 609−611.

Ferris,F.G., Beveridge,T.J. and Fyfe,W.S. (1987a) Manganese oxide deposition in a hot spring microbial mat. *Geomicrobiology Journal* **5**, 33−42.

Ferris,F.G., Beveridge,T.J. and Fyfe,W.S. (1987b) Bacteria as nucleation sites for authigenic minerals in a metal-contaminated lake sediment. *Chemical Geology* **63**, 225−232.

Ferris,F.G., Beveridge,T.J. and Fyfe,W.S. (1987c) Metallic ion binding by *Bacillus subtilis:* implications for the fossilization of microorganisms. *Geology* **16**, 149−152.

Golubic,S. and Barghoorn,E.S. (1977) Interpretation of microbial fossils with special reference to the Precambrian. In: *Fossil Algae* (Ed. E.Flugel), Springer-Verlag, Berlin, pp. 1−14.

Hoyle,B. and Beveridge,T.J. (1983) Binding of metallic ions to the outer membrane of *Escherichia coli*. *Applied and Environmental Microbiology* **46**, 749−752.

Hoyle,B. and Beveridge,T.J. (1984) Metal binding by the peptidoglycan sacculus of *Escherichia coli* K-12. *Canadian Journal of Microbiology* **30**, 204−211.

Leive,L. (1965) Release of lipopolysaccharide by EDTA treatment of *E.coli*. *Biochemical and Biophysical Research Communications* **21**, 290−296.

Martin,N.L. and Beveridge,T.J. (1986) Gentamicin interaction with *Pseudomonas aeruginosa* cell envelope. *Antimicrobial Agents and Chemotherapy* **29**, 1079−1087.

Piper,D.Z., Basler,J.R. and Bischoff,L.L. (1984) Oxidation state of marine manganese nodules. *Geochimica et Cosmochimica Acta* **48**, 2347−2355.

Sellmann,P.V. (1972) Geology and properties of materials exposed in the USA/CRREL permafrost tunnel. *Cold Regions Research and Engineering Laboratory Special Report* **177**, 1−16.

Shaw,P.J., Hills,G.J., Hendwood,J.A., Harris,J.E. and Archer,D.A. (1985) Three-dimensional architecture of the cell sheath and septa of *Methanospirillum hungatei*. *Journal of Bacteriology* **161**, 750−757.

Sonnenfeld,E.M., Beveridge,T.J. and Doyle,R.J. (1985b) Discontinuity of charge on cell wall poles of *Bacillus* on the cell wall of *Bacillus subtilis*. *Journal of Bacteriology* **163**, 1167−1171.

Stewart,M., Beveridge,T.J., and Sprott,G.D. (1985) Crystalline order to high resolution in the sheath of *subtilis*. *Canadian Journal of Microbiology* **31**, 875−877.

Southgate,P.N. (1986) Depositional environment and mechanism of preservation of microfossils, upper Proterozoic Bitter Springs Formation, Australia. *Geology* **14**, 683−686.

Sprott,G.D., Colvin,J.R. and McKellar,R.C. (1979) Sphaeroplasts of *Methanospirillum hungatei* formed upon treatment with dithiothreitol. *Canadian Journal of Microbiology* **25**, 730−738.

Sprott,G.D., Shaw,K.M. and Beveridge,T.J. (1987) Properties of the particulate enzyme $F_{420}$-reducing hydrogenase isolated from *Methanospirillum hungatei*. *Canadian Journal of Microbiology* **33**, 896−904.

Stewart,M., Beveridge,T.J. and Sprott,G.D. (1985) Crystalline order to high resolution in the sheath of *Methanospirillum hungatei*: a cross-beta structure. *Journal of Molecular Biology* **183**, 509−512.

Takano,T. and Dickerson,R.E. (1981) Conformation changes of cytochrome *c*. *Journal of Molecular Biology* **153**, 79−94.

Trudinger,P.A. and Swaine,D.J. (1979) *Biogeochemical Cycling of Mineral-Forming Elements*. Elsevier, Amsterdam.

Turner,S. and Buseck,P.R. (1981) Todorokites: a new family of naturally occurring manganese oxides. *Science* **212**, 1024−1027.

Walker,S.G. and Beveridge,T.J. (1988) Amikacin disrupts the cell envelope of *Pseudomonas aeruginosa* ATCC 9072. *Canadian Journal of Microbiology* **34**, 12−18.

Zeikus,J.G. and Bowen,V.G. (1975) Fine structure of *Methanospirillum hungatei*. *Journal of Bacteriology* **121**, 373−380.

CHAPTER 6

# Biomineralization by magnetogenic bacteria

RICHARD P.BLAKEMORE[1] and RICHARD B.FRANKEL[2]

*[1]Department of Microbiology, University of New Hampshire, Durham, NH 03824, USA and [2]Department of Physics, California Polytechnic State University, San Luis Obispo, CA 93407, USA*

## Introduction

As indicated by the contributions to this symposium, microbes and metals interact in a multitude of ways, in which the latter may have roles ranging from being vital structural and functional cell components to being toxic agents. The magnetotactic bacteria (Blakemore, 1975, 1982; Frankel, 1982) transform extracellular iron into the mixed-valence iron oxide mineral magnetite ($Fe_3O_4$) comprising an intracellular magnetic navigational apparatus (*Figure 1*); they provide a fascinating example of the seemingly unlimited capacity of life to adapt to the physical and chemical world. Readers interested in the role of biogenic magnetite in magnetotaxis and cell navigation in the magnetotactic bacteria and other organisms are referred to several reviews (Frankel, 1982, 1984; Blakemore, 1982; Blakemore *et al.*, 1988).

A purpose of this chapter is to give further emphasis to the poorly characterized physiological group of bacteria, the 'dissimilatory iron-reducers' (a term first suggested in a footnote on page 190 of Ehrlich, 1981; see also Short and Blakemore, 1986; Lovley, 1987) by reviewing aspects of those members that produce magnetite as a biomineralization product. The term 'magnetogenic' bacteria is proposed for these dissimilatory iron reducers. As is apparent from study of their physiology, occurrence and distribution, the magnetotactic bacteria appear to be (facultative) dissimilatory iron-reducers. Recently, dissimilatory iron-reducing bacteria which produce extracellular magnetite have been reported (Lovley *et al.*, 1987; Bell *et al.*, 1987). Each of these types of magnetogens carries out a different manner of biomineralization, producing different forms of the same mineral. Bacterial magnetite, especially as being a biomineralization product of these bacteria and a mineral of biogeochemical interest, is examined in more detail in the pages which follow. Recent reviews by Lovley (1987) and Jones (1986) provide a more comprehensive treatment of the broader subject of dissimilatory iron reduction by microorganisms.

## Prokaryotes and respiratory diversity

Diversity is a hallmark of prokaryotes. As with eukaryotes, this is evident from their great variation in structure and morphology. But it is in respect to their physiological diversity that bacteria simply outclass the rest of life. The bacterium *Escherichia coli*

85

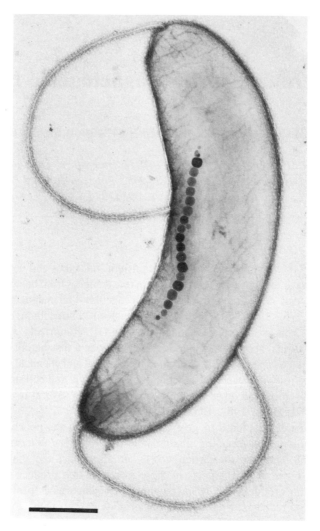

**Figure 1.** A freshwater magnetotactic bacterium illustrative of the BOB-type magnetogen. Intracellular magnetosomes comprise a navigational apparatus causing the cell to align in, and swim along, the geomagnetic field lines. The magnetosomes are of cubo-octahedral morphology. Transmission electron micrograph. The bar represents 0.5 $\mu$m.

can carry out fermentation or respiration of carbohydrates, polyols, carboxylic acids, fatty acids or amino acids. Its respiratory mechanisms (which, as readers interested in metal−microbe interactions are reminded, depend upon a variety of metalloenzymes and metal-centred redox carriers) allow disposal of electrons to oxygen, nitrate, nitrite, fumarate, trimethylamine oxide, dimethylsulphoxide or tetrathionate as the terminal oxidant (Ingledew and Poole, 1984). Less well-studied bacteria oxidize substrates with reduction of sulphate, carbonate or metals, including $Mn^{+4}$ or $Fe^{3+}$. Myers and Nealson (1988) recently described *Alteromonas putrefaciens* strain MR-1 which uses any of the acceptors mentioned above in addition to thiosulfate, sulphite or glycine!

Such catabolic versatility is possible because, unlike mitochondria, prokaryotes may have branched respiratory chains. In response to changes in prevailing environmental conditions it is not uncommon for them to selectively synthesize different oxidoreductase 'modules' (Stewart, 1988; Ingledew and Poole, 1985; Hackett and Bragg, 1983) appropriate for electron disposal to alternate terminal acceptors. Each mode of electron disposal may require a different enzymic pathway, support growth to a different degree and result in distinctive products. Respiratory diversity thus serves to distinguish the physiological groups of prokaryotes: the denitrifiers, sulphate-reducers, methanogens, and the like.

## Respiratory diversity and distribution of bacterial groups

An ecological consequence of prokaryote respiratory diversity is especially noticeable with bacteria inhabiting organic-rich sediments. Such aquatic sediments or waterlogged soils characteristically show vertical stratification into biogeochemical zones. Each zone, dominated by a particular respiratory group of bacteria, is situated within a more or less thermodynamically predictable vertical continuum in the sediment column (Jones, 1980; Sørensen *et al.*, 1979; Karlin *et al.*, 1987; Lovley and Phillips, 1986; Tugel *et al.*, 1986). Physiological types with less efficient respiratory activities occur in progressively deeper strata. Oxygen respirers, having the most energetically favourable respiration (the $\Delta G^0$ for acetate dissimilation with oxygen is $-893.7$ kJ per reaction), predominate in the aerobic zone at the sediment surface. Denitrifiers ($\Delta G^0$ for acetate dissimilation to carbon dioxide with nitrate reduction to nitrite is $-597.4$ kJ per reaction) and 'nitrate ammoniafiers' (Sørensen, 1987) able to exploit a $\Delta G^0$ for acetate dissimilation of $-861.8$ kJ per reaction, are found just below in the microaerobic to anaerobic zone frequently coinciding with the 'redoxcline' (Hallberg, 1978). This transition zone, having an $E'_h$ of approximately $-200$ mV, is often recognizable by the colour change associated with the $Fe^{3+}/Fe^{2+}$ transition. Below this, particularly in marine sediments, is the zone of most active sulphate reduction dominated by the anaerobic sulphate reducers ($\Delta G^0$ for acetate oxidation to carbon dioxide with sulphate reduction is $-161.0$ kJ per reaction). Finally, methanogens dissimilating acetate, for instance, to methane and carbon dioxide ($\Delta G^0$ $-75.7$ kJ per reaction) comprise the predominant physiological group in deepest anoxic sediments.

During early organic matter mineralization and sediment diagenesis the diverse bacterial respiratory groups also participate in a temporal continuum of activity. Until oxygen has been sufficiently depleted, the activity of oxygen respirers at first exceeds then gives way to that of denitrifiers and nitrate ammoniafiers. These aerobes, facultative anaerobes and microaerophiles are, in turn, commonly succeeded by obligate anaerobes, the sulphate reducers, and finally by methanogens.

These predictions concerning microbial spatial and temporal distributions are idealized. They ignore disturbances caused, for example, by bioturbation and introduction of pollutants. They probably result from many factors influencing microbial competitive efficiency including availability of the different oxidants, availability and competitive efficiency for hydrogen, and toxicity of respiratory products of one group (such as nitrite or hydrogen sulphide) towards others. An end result, however, is that organic matter is dissimilated efficiently through concerted attack by members of a spectrum of respiratory types.

Early descriptions of biogeochemical zonation gave relatively little significance to bacteria that use metals (manganese and iron, principally) as oxidants. Research in recent years has increased our appreciation of this aspect of metal – microbe interactions and opened a new chapter in our understanding of metals as agents of organic matter turnover.

## Dissimilatory iron reduction

The $\Delta G^0$ for acetate dissimilation with ferric iron is $-716.3$ kJ per reaction, suggesting that 'iron respiration' (Short and Blakemore, 1986) should drive decomposition of organic matter. As has been pointed out by others, the similarity in $E_0'$ values of the $Fe^{3+}/Fe^{2+}$ couple ($+772$ mV) and the $O_2/H_2O$ couple ($+818$ mV) argues that use of iron as an oxidant should provide for energy conservation efficiencies and molar growth yields with iron not very different from those obtainable with oxygen (Jones, 1982; Thauer *et al.*, 1977).

The importance of thermodynamics notwithstanding, bacterial physiological activity is more obviously responsive to chemical kinetics; the concentration, biological availability and rate of utilization of an oxidant are of more immediate and crucial importance to a metabolizing cell than is its thermodynamic value. Iron being the fourth most abundant crustal element, the ferric iron pool within sediments is a large one. Even though it is stable there only as insoluble ferric oxyhydroxides, phosphates or carbonates, many species of microbes possess adaptive strategies for obtaining this metal from many of these 'unavailable' precipitated minerals. Consequently, for kinetic as well as thermodynamic reasons, dissimilatory iron-reducing bacteria able to couple substrate oxidation with ferric ion reduction should be readily isolated from sediments rich in organic matter. In fact, progress in this area has been only modest. Several reasons for lack of progress comparable to that associated with other respiratory groups can be suggested. First, there are inherent difficulties in working with $Fe^{3+}/Fe^{2+}$ systems at neutral pH in the presence of organic matter (or oxidizing species such as nitrite) because of the ease of chemical reoxidation of the ferrous ions produced. Secondly, in contrasting the quantity of oxidant required to drive organic matter mineralization, the one-electron change from ferric to ferrous ion is less impressive than is, for example, the eight-electron change involved in nitrate reduction to ammonia. Iron respiration requires an abundant supply of ferric ions, which, for the microbiologist, may mean working with an insoluble precipitated phase in the culture medium. Thirdly, dissimilatory iron reducers might commonly be facultative in this regard, showing even more obvious effects of the use of other acceptors as is commonly observed when nitrate is present. Progress is likely to improve. Jones *et al.* (1984) isolated an iron-reducing *Vibrio* species and demonstrated a shift in its metabolic products with Fe(III) added to the medium. The interesting iron-reducing *Alteromonas* strain isolated by Myers and Nealson (1988) was mentioned previously in reference to its metabolic diversity. Lovley and Phillips (1987) have shown that dissimilatory iron-reducing bacteria can outcompete sulphate-reducers and methanogens for needed hydrogen and thereby inhibit the latter in the zone of iron reduction. Such studies can be expected to focus more attention on metal-reducers as agents of biogeochemical importance.

If the thermodynamic generalizations discussed above have predictive value, then the dissimilatory iron-reducers should predominate in the redoxcline, being situated near the zone of most active denitrification and nitrate ammoniafication but above that of most active sulphate reduction. One might also expect that in recently disturbed or freshly collected sediment samples the population of dissimilatory iron-reducers would increase after aerobic and facultatively anaerobic microbial activity involving use of oxygen and nitrate by resident oxygen respirers, denitrifiers and nitrate ammoniafiers. These predictions describe the distribution and activity of magnetotactic bacteria and, perhaps, those of other magnetogenic bacteria as well.

## BIM, BOB and biogenic magnetite

Magnetite is a common biomineral produced by both prokaryotes and eukaryotes (Lowenstam, 1981). It can be produced, even among magnetogenic iron-reducing bacteria, by either of the two fundamentally different biomineralization processes which have been distinguished (Lowenstam, 1981; Mann, 1986, 1988). These two biomineralization processes differ in significant ways: for mineral formation as a result of biological activity but without biological control, the term 'biologically-induced' mineralization (BIM) has been advanced. BIM occurs as a consequence of the effect of organisms in modifying the chemical composition, pH and redox conditions in the local extracellular environment thereby bringing about chemical precipitation. 'Biologically controlled biomineralization', on the other hand suggests what the term imples. Biominerals produced this way are formed either in association with an organic support structure ('organic matrix-mediated biomineralization; Lowenstam, 1981) or in association with a biological surface such as a membrane ('boundary-organized biomineralization; Mann, 1986). Boundary-organized biomineralization (BOB) occurs in response to spatial localization of chemical precipitation reactions and has other features not shared by BIM. The most important consequence is the extent of biological control over the mineral-forming process provided by BOB (Mann, 1986). As mentioned, the mixed-valence iron oxide magnetite ($Fe_3O_4$) is a biomineral which can be produced by either BIM or BOB processes. A chemical feature common to either process of magnetogenesis is iron reduction. However, the process type, the site of iron reduction and location of the final product (intracellular or extracellular), depend upon the bacterial species involved.

## Magnetite-producing bacteria: the 'magnetogens'

Until recently, the only known magnetite-forming ('magnetogenic') bacteria were the magnetotactic bacteria (Blakemore, 1975, 1982). This situation has now changed markedly with the isolation of bacteria from river and creek sediment which produce extracellular magnetite as a result of ferric iron reduction (Lovley *et al.*, 1987; Bell *et al.*, 1987). One isolate, GS-15 (Lovley *et al.*, 1987) is the first organism known to couple reduction of iron in hydrous ferric oxide to anaerobic acetate oxidation. The ferrous ions produced (eight moles per mole acetate utilized) were kept outside the cell where they reacted with unreacted hydrous ferric oxide to produce large amounts of extracellular ultrafine grained magnetite by a BIM process (Frankel, 1987). Little information is formally available as yet concerning taxonomy and physiology of this Gram-negative BIM-type magnetogen.

Bell *et al.* (1987) obtained a number of Gram-negative, anaerobic, rod-shaped bacteria as isolates from iron-rich creek water. These were magnetogenic when incubated collectively in various combinations but not in axenic monocultures. Ferrous ions were released into culture media containing ferric oxyhydroxides. Except when the pH dropped in response to glucose added to the medium, extracellular magnetite was formed by means of a BIM process. These workers suggested that on the basis of calculations of appropriate $E'_h$ −pH relationships and the usual concentrations of available sulphide, activities of magnetogenic bacteria would not be expected to control organic matter turnover in at least some riverine sediments. They correctly call for more study of dissimilatory iron reduction in anaerobic habitats.

Most successful efforts to locate and collect magnetotactic bacteria result from sampling stabilized sediments at, or near, the redoxcline. This $Fe^{3+}/Fe^{2+}$ transition zone usually coincides with the sediment−water interface of aged laboratory enrichment cultures (Blakemore, 1982). A variety of magnetotactic bacteria frequently achieve very high population densities ($\geq 10^6$ ml$^{-1}$) in this region. Moreover, they usually enrich in mud samples after an initial period of high biological activity followed by gradual clearing of other bacteria from the overlying water. This could be associated with an ability to continue with iron as oxidant after depletion of the more thermodynamically suitable electron acceptors oxygen and nitrate. Thus, in respect to both when and where they preferentially accumulate in mud enrichment cultures, the magnetotactic bacteria behave in a manner expected of dissimilatory iron-reducers. It is incidentally noteworthy that despite the use of anaerobic techniques, and recent success in isolating anaerobic magnetite-producing bacteria (see below), no reports have yet been made of fermentative or sulphate-reducing or methanogenic, magnetogenic bacteria. Magnetogenesis may, for some unknown reason, be restricted to organisms which respire using oxidants of higher electronegativity such as oxygen, nitrate, nitrous oxide or $Fe^{3+}$.

The obligately microaerophilic, denitrifying bacterium *Aquaspirillum magnetotacticum* strain MS-1 (Blakemore *et al.*, 1979) uses as terminal electron acceptors, oxygen, nitrate (Bazylinski and Blakemore, 1983) and, apparently, $Fe^{3+}$. Because it will not grow anaerobically with any acceptor yet tested, and apparently because of complexities associated with metal precipitation and chemical reoxidation of the $Fe^{2+}$ formed, it has proved difficult to critically evaluate a molar growth yield with $Fe^{3+}$ (W.Guerin, unpublished observations). Nevertheless, *A.magnetotacticum* cells actively reduce iron when grown at very high (2 mM) concentrations of iron as ferric oxyhydroxides. In phosphate-buffered medium, for instance, growing cells of strain MS-1 produced (as a result of a BIM process), significant amounts of the reduction product, vivianite [$Fe_3(PO_4)_2$] which precipitated as needle-like extracellular crystals (*Figure 2*) in the culture medium (N.Blakemore, R.Frankel and W.Guerin, unpublished observations). Strain MS-1 is also capable of 'iron respiration' (Short and Blakemore, 1986). When provided with ferric iron anaerobically, intact cells translocated protons outwardly in direct demonstration of its role as a terminal electron acceptor (Short and Blakemore, 1986).

Iron reductase activity assayed in cell fractions was located predominantly in the periplasm (77% of total reductase detected) with only 3% in the cell membranes (Paoletti and Blakemore, 1988). Iron reduction in the periplasm is probably very important in its transport into the cell and, consequently, may represent a facet of assimilatory iron reduction by this organism. The relationship of pathways of iron assimilation and those of iron dissimilation to magnetite formation are incompletely understood at present.

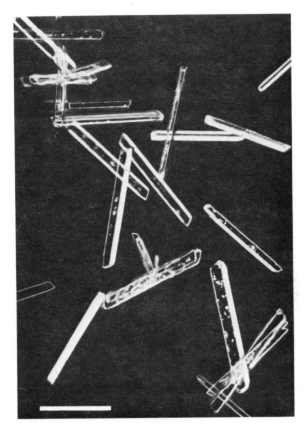

**Figure 2.** Extracellular crystals of vivianite [$Fe_3(PO_4)_2$] produced as a result of a BIM process by cells of *Aquaspirillum magnetotacticum* in phosphate-buffered medium containing high iron. Darkfield optical micrograph. The bar indicates 25 $\mu$m.

In terms of iron dissimilation, a periplasmic location of respiratory iron reductase would (i) obviate the potential membrane barrier presented to ferric iron were its terminal reduction sites intracellular and (ii) promote a favourable proton gradient by keeping the protons that accumulate during iron reduction outside the cell. However, the insensitivity to several respiratory inhibitors of the bulk iron reductase activity detected in strain MS-1 did not appear consistent with a function in cell respiration under the conditions investigated (Paoletti and Blakemore, 1988). Similar findings for *E.coli* K-12 were obtained by Williams and Poole (1987). The major reductase activity measured in *A.magnetotacticum* was produced constitutively. Respiratory iron reductase may comprise only a fraction of the total activity detected and require induction by, as yet, unknown factors.

Although not yet tested exhaustively, in one study cell ability to respire with iron correlated with an ability to produce magnetite; intact cells of a non-magnetic mutant strain failed to eject protons when provided with $Fe^{3+}$ (Short and Blakemore, 1986). Studies aimed at determining possible relationships between formation of intracellular magnetite by magnetotactic bacteria and their use of iron as oxidant remain inconclusive.

Recent success with chemostat culture of this organism can be expected to clarify such relationships with cells grown to high cell yield under reproducible and controlled conditions. In such circumstances, intracellular magnetite production by this organism can now be precisely regulated by culture oxygen or iron supply (Y.Gorby, unpublished data) in anticipation of studies aimed at possible gene regulation of magnetogenesis.

## Formation and characteristics of biogenic magnetite

Magnetite is distinguished among many iron minerals by its X-ray or electron diffraction spectra for which amounts of material as little as one crystal are needed. Magnetite also has a characteristic iron-57 Mössbauer resonance spectrum (Frankel *et al.*, 1979). This spectroscopic technique is non-destructive to the sample and is unaffected by associated biological or non-ferrous material. It provides information concerning valence and chemical form of iron (iron−sulphur compounds are distinguished from those having iron in oxygen coordination) in biological samples or in sediments. Results of an extensive iron-57 Mössbauer spectroscopic study of *A.magnetotacticum* revealed the nature and distribution of the iron compounds present within cells and cell fractions. By comparing the major iron compounds present in wild-type and mutant cells as well as those within cells examined at different stages of growth, evidence was obtained that magnetite was formed through reduction of one-third of the $Fe^{3+}$ in the high-density ferric oxide precursor, ferrihydrite (Frankel *et al.*, 1983). Reduction of iron in ferrihydrite within cores of mammalian ferritin does not produce magnetite. Thus, iron reduction is necessary but insufficient to ensure magnetite formation. How can crystals of magnetite be deposited as 'magnetosomes' (Balkwill *et al.*, 1980) having a particular (non-random) distribution and orientation within the magnetotactic bacteria?

Gorby *et al.* (1988) showed that the magnetosomes of *A.magnetotacticum* strain MS-1 were each within a vesicle enclosed by a biological membrane consisting of a lipid bilayer. This appeared also to be the case with magnetosomes within other species of magnetotactic bacteria (Mann *et al.*, 1987a). The mechanism whereby the cell precipitates the inverse spinel crystal structure of magnetite and not the hexagonal close-packed structure of hematite ($\alpha$-$Fe_2O_3$) remains unknown. However, it is only at low $E_h'$ and high pH that magnetite is the preferred phase in an iron−oxygen−water system, and at pH 7 is is stable only at $E_h'$ values less than $-100$ mV (Garrels and Christ, 1965; see also Bell *et al.*, 1987). Membranous vesicles within the cell could allow for compartmentalization of its cytoplasm and allow, in turn, for segregation of localized intracellular sites of appropriate low $E_h'$, high pH and supersaturating ion concentrations required for magnetite formation. Evidence of compartmentalization of cell ion pools in a magnetotactic bacterium is shown in *Figure 3*. This type of organism has been shown by means of energy dispersive X-ray analysis to segregate phosphorus and potassium (which map to the large, dense spherical structures shown) from iron, which maps only in the position occupied by magnetosomes (unpublished results). Mann *et al.* (1984a, 1987b) have proposed that intracellular magnetite results from a process involving precipitation, solution and re-precipitation of ions within vesicles, the boundary membrane of which is imagined to exert a major influence on crystal dimensions, morphology and cell location.

The magnetosome membrane of *A.magnetotacticum* contains 20, or more, proteins, several of which are not found in other cell membranes (Gorby *et al.*, 1988). These

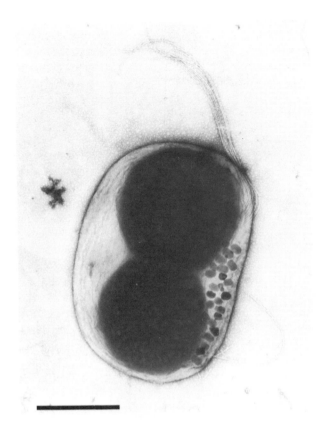

**Figure 3.** A freshwater magnetotactic bacterium containing magnetosomes of hexagonally prismatic morphology. Note the compartmentalization of the cell into regions containing phosphorus and potassium (dense, spherical structures) distinct from regions containing iron (magnetosomes). Transmission electron micrograph. The bar indicates 1 μm.

latter may function in reductase or dehydratase activities important for formation of magnetite from the hydrous ferric oxide precursors detected by Mössbauer spectroscopy. It is likely that, as a part of the BOB process in the magnetotactic bacteria, the magnetosome membrane and its components play roles in partitioning off cytoplasm, supersaturating the compartment so formed with appropriate ions, and determining the maximum size and distribution of magnetite grains formed within the cell. In addition, the membrane or its components may control the magnetite crystal morphology.

Crystal lattice imaging studies of bacterial magnetosomes by means of high-resolution transmission electron microscopy (HRTEM) have revealed several distinctive crystal forms of magnetite within magnetotactic bacteria. All are single (or occasionally twinned) crystals. The type found within *A.magnetotacticum* and other cells (*Figure 1*) which appears in horizontal projection to be rectangular, consists of truncated octahedral prisms (Mann *et al.*, 1984a). Others, which appear in horizontal projection as rectangular parallelepipeds (*Figure 3*), consist of truncated hexagonal prisms (Mann *et al.*, 1984b; Matsuda *et al.*, 1983). Those of a third type, which appear tapered, curved, or

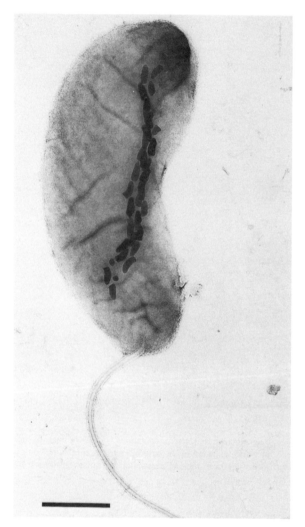

**Figure 4.** A freshwater magnetotactic bacterium containing magnetosomes of the anisotropic type. Transmission electron micrograph. The bar indicates 0.5 µm.

bullet-shaped in horizontal projection (*Figure 4*), are anisotropic but based upon a hexagonally prismatic structure (Mann *et al.*, 1987b). Mann *et al.* (1984a, 1987b) have suggested that variation in crystal morphology could result from isotropic single crystals with a common cubo-octahedral morphology, which experience subsequent growth along selected axes and suppression of growth along others. Selective growth might be controlled by the membrane. The crystal morphologies appear to be species-specific. All cells having a particular and distinctive morphology always possess magnetosomes of particular crystal morphology (see *Figures 1, 3* and *4*). Cells of *A.magnetotacticum* strain MS-1 always produce octahedral particles regardless of changes in the culture medium or growth conditions. Thus, determination of the crystal morphology appears to have a strong genetic component, although too few species of magnetotactic bacteria have been grown in pure culture to evaluate this. Crystals with morphologies of several of the types observed within magnetotactic bacteria have not been produced by other than biogenic means. Biogenic minerals produced by BIM possess shapes characteristic of the chemically precipitated mineral. Those produced by BOB may have unique and distinct morphology different from the chemically precipitated forms of the same mineral (Lowenstam, 1981). This appears true of biogenic magnetite. Crystals produced by the BIM-type magnetogen GS-15 lack the unique morphologies of magnetosomes within the BOB-type magnetogens, the magnetotactic bacteria. This is an important factor in evaluating the possible presence and origin of biogenic magnetite in natural habitats as discussed below.

Analysis of bacterial magnetite, be it produced by BIM or BOB processes, has revealed several other interesting properties. The most important of these is its single magnetic domain (SD) character. Crytals of magnetite in the SD size range are roughly from 400 to 1000 Å along a major axis (Butler and Banerjee, 1975; Frankel, 1981). Consequently, it is possible to evaluate this property of bacterial magnetite by means of electron microscopic measurement. However, the SD character of ultrafine grain magnetic particles can be directly measured with a sensitive magnetometer known as a superconducting quantum interference device (SQUID). Thus, it is not necessary to view the SD particles in order to determine their presence. This enables direct examination of sediments without the need for sample clean-up as for electron microscopy. To distinguish SD from multidomain (MD) particles the Lowrie – Fuller test (Lowrie and Fuller, 1971) is applied. Typically, the resistance to demagnetization with an alternating field (AF) is measured. The (weak field) anhysteritic remanent magnetization (ARM) of SD and small MD particles resists AF demagnetization more than does (strong field) saturation isothermal remanent magnetization (sIRM). Large MD particles have opposite behaviour. Thus, if the median destructive field of ARM is greater than that of sIRM of a sample, then it contains predominantly SD magnetic particles. These non-invasive magnetic measurements can be made without destroying a sample and with no perturbation other than exposing it to magnetizing and demagnetizing fields. In each of the dozens of types of magnetotactic bacteria which have been examined, the particles are SD.

The extracellular magnetite precipitated by BIM through the action of the bacterium GS-15 also contains SD but not MD grains as evidenced from the Lowrie – Fuller test (Lovley *et al.*, 1987b). More recent data, however, (R.B.Frankel, unpublished) indicated that although SD particles were present, the major amount of magnetite produced by GS-15 in laboratory culture consisted of superparamagnetic grains smaller than SD. Moreover, it is different from the highly regular and homogeneous grains with distinctive morphologies that comprise magnetosomes.

## Biogenic magnetite in sediments and rocks

Kirschvink and Chang (1984) and Stolz *et al.* (1986) investigated the possibility, first suggested by Kirschvink and Lowenstam (1979), that biogenic magnetite might persist in deep-sea sediments and thereby contribute a biological component to the paleomagnetic record. In their extracted fine-grain magnetic fractions of sediments they observed SD magnetite crystals with the morphologies peculiar to those of bacterial magnetosomes.

They proposed that these were of bacterial origin. Petersen *et al.* (1986), in characterizing magnetic phases in South Atlantic deep-sea sediments, observed a predominance of highly homogeneous SD magnetite in most samples studied. They carried out improved magnetic separation techniques to recover SD magnetite from sediment and applied electron microscopic analysis to the separated fractions. Their results (see especially von Dobeneck *et al.*, 1987) convincingly reveal the similarities of deep-sea SD magnetite grains to those they also found within magnetotactic bacteria and clearly different from magnetite grains of lithogenic origin. Additional crystal morphologies and arrangements of bacterial magnetosomes produced by BOB-type magnetogens may be found in Vali *et al.* (1987) and in Sparks *et al.* (1986).

Recently Karlin *et al.* (1987) examined the magnetic and geochemical properties of suboxic hemipelagic marine sediments. They obtained evidence for the formation of SD magnetite in (the brown-tan-green) $Fe^{3+}/Fe^{2+}$ transition zone, more specifically in a region between the zone of nitrate reduction and that of iron reduction. They attributed its presence to the activity of magnetotactic bacteria there.

In laboratory experiments, *A.magnetotacticum* strain MS-1 cells did not grow anaerobically and formed optimal quantities of magnetite under microaerobic conditions, that is, with the dissolved oxygen tension of the culture at approximately 1% of saturation (Blakemore *et al.*, 1985). At the proposed location, magnetogenic organisms would be expected to be anaerobic. Thus, organism GS-15 would be a better candidate as a biogenic source of magnetite than would *A.magnetotacticum* strains MS-1. Karlin *et al.* (1987) did not report information concerning magnetite crystal morphology which would help resolve its origin, as noted above. Of great significance in obtaining a more complete understanding of the origin of sedimentary magnetite, is the recent isolation in Woods Hole, Massachusetts, of a marine, magnetotactic bacterium MV-1 (Bazylinski *et al.*, 1988). Unlike *A.magnetotacticum* strain MS-1, this organism produces intracellular hexagonally prismatic magnetosomes under anaerobic conditions. It is no longer necessary to eliminate the magnetotactic bacteria from consideration as potential sources of biogenic magnetite in anoxic marine habitats.

Metals including iron are assuming more prominent roles in our understanding of sediment diagenesis and organic matter mineralization in aquatic habitats. One reason is the growing success among microbiologists in isolating microbes responsible for metal transformations. As pointed out in this chapter, both the BIM-type magnetogens and BOB-type magnetogens provide examples of groups with which recent progress has been exciting. Equally significant is the increased coming together of the required expertise provided by ecologists, physiologists, biogeochemists, geologists and inorganic chemists and their different armamentaria. Such concerted attack will be important in filling the potentially difficult remaining gaps in our knowledge of biomineralization processes in the biogeochemical cycling of metals and turnover of organic matter.

## Acknowledgements

R.P.B. thanks the Society for General Microbiology and Cell Biology Group for inviting and supporting his participation in this symposium. Assistance of N.Blakemore in sample collection, microscopy and preparation of the resulting figures is gratefully acknowledged. Work carried out in the laboratory of R.P.B. is supported by the National

Science Foundation grant DMB 85-15540 and Office of Naval Research contract N00014-85-K-0502. R.B.F. acknowledges support of Office of Naval Research contract N00014-85-K-0505.

## References

Balkwill,D.L., Maratea,D. and Blakemore,R.P. (1980) Ultrastructure of a magnetotactic spirillum. *Journal of Bacteriology* **141**, 1399–1408.

Bazylinski,D.A. and Blakemore,R.P. (1983) Denitrification and assimilatory nitrate reduction in *Aquaspirillum magnetotacticum. Applied and Environmental Microbiology* **46**, 1118–1124.

Bazylinski,D.A., Frankel,R.B. and Jannasch,H.W. (1988) Anaerobic magnetite production by a marine, magnetotactic bacterium. *Nature (London)* **334**, 518–519.

Bell,P.E., Mills,A.L. and Herman,J.S. (1987) Biogeochemical conditions favoring magnetite formation during anaerobic iron reduction. *Applied and Environmental Microbiology* **53**, 2610–2616.

Blakemore,R.P. (1975) Magnetotactic bacteria. *Science* **190**, 377–379.

Blakemore,R.P. (1982) Magnetotactic bacteria. *Annual Review of Microbiology* **36**, 217–238.

Blakemore,R.P., Blakemore,N.A. and Frankel,R.B. (1988) Bacterial biomagnetism and geomagnetic field detection by organisms. In *Modern Bioelectricity* (ed. A.A.Marino), Marcel Dekker, New York, pp. 19–34.

Blakemore,R.P., Maratea,D. and Wolfe,R.S. (1979) Isolation and pure culture of a freshwater magnetic spirillum in chemically defined medium. *Journal of Bacteriology* **140**, 720–729.

Blakemore,R.P., Short,K.A., Bazylinski,D.A., Rosenblatt,C. and Frankel,R.B. (1985) Microaerobic conditions are required for magnetite formation within *Aquaspirillum magnetotacticum. Geomicrobiology Journal* **4**, 53–71.

Butler,R.F. and Banerjee,S.K. (1975) Theoretical single-domain grain size range in magnetite and titanomagnetite. *Journal of Geophysical Research* **80**, 4049–4058.

Ehrlich,H.L. (1981) *Geomicrobiology.* Marcel Dekker, New York.

Frankel,R.B. (1982) Magnetotactic bacteria. *Comments in Molecular and Cellular Biophysics* **1**, 293–310.

Frankel,R.B. (1984) Magnetic guidance of organisms. *Annual Review of Biophysics and Bioengineering* **13**, 85–103.

Frankel,R.B. (1987) Anaerobes pumping iron. *Nature (London)* **330**, 208.

Frankel,R.B., Blakemore,R.P. and Wolfe,R.S. (1979) Magnetite in freshwater magnetotactic bacteria. *Science* **203**, 1355–1356.

Frankel,R.B., Papaefthymiou,G.C., Blakemore,R.P. and O'Brien,W. (1983) $Fe_3O_4$ precipitation in magnetotactic bacteria. *Biochimica et Biophysica Acta* **763**, 147–159.

Garrels,R.M. and Christ,C.L. (1965) *Solutions, Minerals and Equilibria.* Freeman, Cooper and Co., San Francisco.

Gorby,Y.A., Beveridge,T.J. and Blakemore,R.P. (1988) Characterization of the bacterial magnetosome membrane. *Journal of Bacteriology* **170**, 834–841.

Hackett,N.R. and Bragg,P.D. (1983) Membrane cytochromes of *Escherichia coli* grown aerobically and anaerobically with nitrate. *Journal of Bacteriology* **154**, 708–718.

Hallberg,R. (1978) Metal–organic interaction at the redoxcline. In: *Environmental Biogeochemistry and Geomicrobiology* (Ed. W.E.Krumbein), Science Ann Arbor, Michigan, 947–953.

Ingeldew,W.J. and Poole,R.K. (1984) The respiratory chains of *Escherichia coli. Microbiological Reviews* **48**, 222–271.

Jones,C.W. (1982) Bacterial respiration and photosynthesis. In *Aspects of Microbiology* 5. American Society for Microbiology, Washington, DC.

Jones,G.E. (1980) Biogeochemical succession of bacterial activities in aquatic sediments. *Microbiology* 1980, 348–349.

Jones,J.G. (1986) Iron transformations by freshwater bacteria. *Advances in Microbial Ecology* **9**, 149–185.

Jones,J.G., Gardener,S. and Simon,B.M. (1984) Reduction of ferric iron by heterotrophic bacteria in lake sediments. *Journal of General Microbiology* **130**, 45–51.

Karlin,R., Lyle,M. and Heath,G.R. (1987) Authigenic magnetite formation in suboxic marine sediments. *Nature (London)* **326**, 490–493.

Kirschvink,J.L. and Chang,S.-B.R. (1984) Ultrafine-grained magnetite in deep-sea sediments: possible bacterial magnetofossils. *Geology* **12**, 559–562.

Kirschvink,J.L. and Lowenstam,H.A. (1979) Mineralization and magnetization of chiton teeth: paleomagnetic, sedimentologic and biologic implications of organic magnetite. *Earth and Planetary Science Letters* **44**, 193−204.

Lovley,D.R. (1987) Organic matter mineralization with reduction of ferric iron: a review. *Geomicrobiology Journal* **5**, 375−399.

Lovley,D.R. and Phillips,E.J.P. (1986) Organic matter mineralization with reduction of ferric iron in anaerobic sediments. *Applied and Environmental Microbiology* **51**, 683−689.

Lovley,D.R. and Phillips,E.J.P. (1987) Competitive mechanisms for inhibition of sulfate reduction and methane production in the zone of ferric iron reduction in sediments. *Applied and Environmental Microbiology* **53**, 2636−2641.

Lovley,D.R., Stolz,J.F., Nord,G.L., Jr and Phillips,E.J.P. (1987) Anaerobic production of magnetite by a dissimilatory iron-reducing microorganism. *Nature (London)* **330**, 252−254.

Lowenstam,H.A. (1981) Minerals formed by organisms. *Science* **211**, 1126−1131.

Lowrie,W. and Fuller,M. (1971) On the alternating field demagnetization characteristics of multidomain thermoremanent magnetization in magnetite. *Journal of Geophysical Research* **76**, 6339−6349.

Mann,S. (1986) On the nature of boundary-organized biomineralization (BOB). *Journal of Inorganic Chemistry* **28**, 363−371.

Mann,S. (1988) Molecular recognition in biomineralization. *Nature (London)* **322**, 119−124.

Mann,S., Frankel,R.B. and Blakemore,R.P. (1984a) Structure, morphology and crystal growth of bacterial magnetite. *Nature (London)* **310**, 405−407.

Mann,S., Moench,T.T. and Williams,R.J.P. (1984b) A high-resolution electron microscopic investigation of bacterial magnetite. Implications for crystal growth. *Proceedings of the Royal Society of London B* **221**, 385−393.

Mann,S., Sparks,N.H.C. and Blakemore,R.P. (1987a) Ultrastructure and characterization of anisotropic magnetic inclusions in magnetotactic bacteria. *Proceedings of the Royal Society of London B* **231**, 469−476.

Mann,S., Sparks,N.H.C. and Blakemore,R.P. (1987b) Structure, morphology and crystal growth of anisotropic magnetite crystals in magnetotactic bacteria. *Proceedings of the Royal Society of London B* **231**, 477−487.

Matsuda,T., Endo,J., Osakabe,N. and Tonomura,A. (1983) Morphology and structure of biogenic magnetite particles. *Nature (London)* **302**, 411−412.

Myers,C.R. and Nealson,K.H. (1988) Bacterial manganese reduction and growth with manganese oxide as the sole electron acceptor. *Science* **240**, 1319−1321.

Paoletti,L.C. and Blakemore,R.P. (1988) Iron reduction by *Aquaspirillum magnetotacticum*. *Current Microbiology* **17**, 339−342.

Petersen,N., von Dobeneck,T. and Vali,H. (1986) Fossil bacterial magnetite in deep-sea sediments from the South Atlantic Ocean. *Nature (London)* **320**, 611−615.

Short,K.A. and Blakemore,R.P. (1986) Iron respiration-driven proton translocation in aerobic bacteria. *Journal of Bacteriology* **167**, 729−731.

Sørensen,J. (1987) Nitrate reduction in marine sediment: pathways and interaction with iron and sulfur cycling. *Geomicrobiology Journal* **5**, 401−421.

Sørensen,J., Jorgensen,B.B. and Revsbech,N.P. (1979) A comparison of oxyen, nitrate and sulfate reduction in coastal marine sediments. *Microbial Ecology* **5**, 105−115.

Sparks,N.H.C., Courtaux,L., Mann,S. and Board,R.G. (1986) Magnetotactic bacteria are widely distributed in sediments in the U.K. *FEMS Microbiology Letters* **37**, 305−308.

Stewart,V. (1988) Nitrate respiration in relation to facultative metabolism in enterobacteria. *Microbiological Reviews* **52**, 190−232.

Stolz,J.F., Chang,S.-B.R. and Kirschvink,J.L. (1986) Magnetotactic bacteria and single-domain magnetite in hemipelagic sediments. *Nature (London)* **321**, 849−851.

Thauer,R.K., Jungermann,K. and Decker,K. (1977) Energy conservation in chemotrophic anaerobic bacteria. *Bacteriological Reviews* **41**, 100−180.

Tugel,J.B., Hines,M.E. and Jones,G.E. (1986) Microbial iron reduction by enrichment cultures isolated from estuarine sediments. *Applied and Environmental Microbiology* **52**, 1167−1172.

Vali,H., Forster,O., Amarantidis,G. and Petersen,N. (1987) Magnetotactic bacteria and their magnetofossils in sediments. *Earth and Planetary Science Letters* **86**, 389−400.

Von Dobeneck,T., Petersen,N. and Vali,H. (1987) Bakterielle magnetofossilien. *Geowissenschaften in unserer Zeit* **5**, 27−35.

Williams,H.D. and Poole,R.K. (1987) Reduction of iron (III) by *Escherichia coli* K-12: lack of involvement of the respiratory chains. *Current Microbiology* **15**, 319−324.

CHAPTER 7

# Mineral-oxidizing bacteria: metal – organism interactions

P.R.NORRIS

*Department of Biological Sciences, University of Warwick, Coventry CV4 7AL, UK*

## Introduction

In relation to metals, the most familiar aspects of the growth of acidophilic, mineral-oxidizing bacteria are probably the oxidation of Fe(II), providing energy for growth, and the capacity of this growth to proceed in the presence of high concentrations of some metals that are potentially toxic. The majority of work concerning both the leaching of minerals by bacteria and how these bacteria interact with metal ions has involved a single species, *Thiobacillus ferrooxidans*, and much of this work has simply described the solubilization of metals and their effects on growth without revealing details of the underlying reactions or processes. An exception with regard to toxic metals has been the characterization of $Hg^{2+}$ reduction by *T. ferrooxidans* which follows similar work with other bacteria.

The physiology and biochemistry of iron oxidation, with *T. ferrooxidans* at least, has attracted attention for over thirty years and thorough reviews have reflected the progress from uncertainty about the route of electron transfer from Fe(II) to oxygen (Tuovinen and Kelly, 1972) to a more confident appraisal of the bioenergetics of the process (Ingledew, 1982). However, our understanding of even this most fundamental aspect of the activity of *T. ferrooxidans* remains incomplete and the process in other acidophiles remains to be studied. The oxidation of some metals other than iron, the Cu(I) in covellite (CuS), for example (see Kelly *et al.*, 1979), is probably necessary for the dissolution of some mineral sulphides. The possibility of direct bacterial oxidation in such cases is difficult to evaluate because Fe(III) both in natural minerals and possibly associated with the surface of iron-oxidizing bacteria could be an intermediate oxidant. The evidence presented for the direct oxidation of $Cu^+$ and $Sn^{2+}$ ions by *T. ferrooxidans* includes manometric and cytochrome reduction data (Lewis and Miller, 1977). In the case of uranium oxidation, the different kinetics of $U^{4+}$ and $Fe^{2+}$ oxidation, and particularly the lack of stimulation by less than 0.5 $\mu$M iron of the slower $U^{4+}$ oxidation, have been cited as evidence of the direct utilization of the uranium as an electron donor (DiSpirito and Tuovinen, 1982). Growth of the bacteria was not obtained, however, with $U^{4+}$ as a potential substrate or, in the earlier work, with $Sn^{2+}$.

The metals required at trace levels by the mineral-oxidizing bacteria have received little attention because they are rarely growth-limiting. The minerals and high concentrations of Fe(II) used as substrates, in otherwise defined media, usually provide sufficient quantities of various trace elements, particularly in view of the low cell yields associated with autotrophic growth on iron. However, there are requirements related specifically to growth on minerals, including those for copper in the rusticyanin of the respiratory chain of iron-oxidizing *T.ferrooxidans* (Cobley and Haddock, 1975), for molybdenum in sulphur oxidation by *Thiobacillus thiooxidans* (Takakuwa *et al.*, 1977) and (see later) for zinc in the iron-oxidizing *Leptospirillum ferrooxidans*.

This chapter summarizes the characteristics of Fe(II) oxidation by *T.ferrooxidans* with the emphasis on how this oxidation and the growth of the bacteria are affected by Fe(III) and other metals. The intention is to illustrate how these interactions compare with those of metals with less familiar mineral-oxidizing bacteria. The other acidophilic bacteria that have received some study in this context fall into three groups. These are (with references to their isolation and growth): the mesophile *L.ferrooxidans* (Balashova *et al.*, 1974; Pivovarova *et al.*, 1981; Norris, 1983); the moderately thermophilic iron-oxidizing bacteria (Norris and Barr, 1985; Wood and Kelly, 1985; Norris *et al.*, 1986; Karavaiko *et al.*, 1988); and the mineral-oxidizing strains of *Sulfolobus* (Brierley and Brierley, 1986; Marsh *et al.*, 1983). Comparisons across these groups of some of the characteristics of the bacteria and their activities in mineral oxidation are also available (Pivovarova and Golovacheva, 1985; Harrison, 1986; Hutchins *et al.*, 1986; Norris, 1988).

## Growth of the iron-oxidizing acidophiles

The doubling times observed for iron-oxidizing *T.ferrooxidans*, *L.ferrooxidans*, moderate thermophiles and *Sulfolobus* strains in batch culture at their respective temperature optima are generally in the range of $5-10$ h. More rapid growth-associated iron oxidation by the moderate thermophiles is possible during chemolitho-heterotrophic growth utilizing yeast extract (Marsh and Norris, 1983) and during mixotrophic growth utilizing glucose, fructose or sucrose (Wood and Kelly, 1985). The autotrophic growth yields of the mesophiles and of the moderate thermophiles are similar in batch culture at about 0.3 g dry weight per g atom $Fe^{2+}$ oxidized, but higher with growth of the thermophiles in the presence of organic nutrients (Kelly and Jones, 1978; Eccleston *et al.*, 1985; Wood and Kelly, 1985). The true growth yield of *T.ferrooxidans* in chemostat culture has been calculated as 1.33 g dry weight per g atom $Fe^{2+}$ oxidized (Kelly *et al.*, 1977). Yield data are not available for *Sulfolobus* strains utilizing Fe(II) at high temperatures where significant abiotic oxidation of the substrate also occurs (Marsh *et al.*, 1983).

It has been suggested that, in most environments, the growth of *T.ferrooxidans* could be limited by product inhibition effects (Jones and Kelly, 1983). Fe(III), therefore, is an important metal to consider in the capacity of an inhibitor as well as of a potential intermediate in the oxidation of other metals and of sulphur.

## The kinetics of iron oxidation

The influence of iron concentration on the growth of the iron-oxidizing acidophiles can be complex, particularly in the chemostat where substrate and different forms of product inhibition of the growth of *T.ferrooxidans* have been observed (Jones and Kelly, 1983). The product inhibition was predominantly competitive or non-competitive and

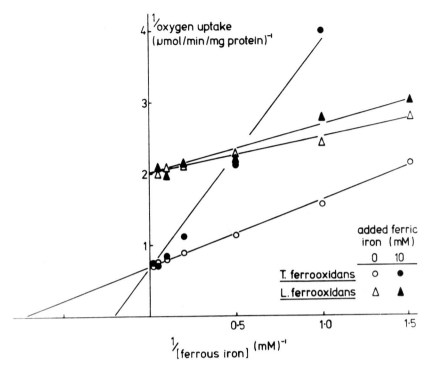

**Figure 1.** Reciprocal plots of oxygen uptake by cell suspensions of mesophilic, iron-oxidizing bacteria at pH 1.7, 30°C and different iron sulphate concentrations. (From Norris *et al.*, 1988.)

**Table 1.** Apparent affinity ($K_m$) and inhibitor ($K_i$) constants (mean, standard deviation and sample size) derived from Lineweaver−Burk plots of data from oxygen electrode experiments with washed cell suspensions of ferrous iron-grown or (in the case of *Sulfolobus* only) pyrite-grown bacteria. (From Norris *et al.*, 1988.)

| Organism | Temp. (°C) | $K_m$ (mM $Fe^{2+}$) | $K_i$ (mM $Fe^{3+}$) |
|---|---|---|---|
| *T. ferrooxidans* | 30 | 1.34 ± 0.16 (6) | 3.10 ± 0.18 (4) |
| *L. ferrooxidans* | 30 | 0.25 ± 0.08 (8) | 42.8 |
| Strain TH1 | 50 | 1.04 ± 0.08 (5) | 2.74 ± 0.13 (3) |
| Strain ALV | 50 | 2.96 ± 0.68 (6) | 1.13 ± 0.34 (5) |
| Strain TH3 | 50 | 0.47 ± 0.10 (4) | 1.89 ± 0.39 (4) |
| *Sulfolobus* BC | 65 | 0.56 ± 0.03 (3) | 1.65 |

affected by various factors, which included the pH and the potassium concentration of the medium. The apparent $K_m$ values, in the range 0.43−0.9 mM Fe(II), which were obtained for iron oxidation by non-growing cell suspensions of *T. ferrooxidans* in an oxygen electrode, were similar to the substrate saturation coefficient ($K_s$) values of 0.7−2.4 mM Fe(II) obtained from chemostat work (Kelly *et al.*, 1977; Kelly and Jones, 1978). The effect of Fe(I) on iron oxidation in the electrodes was simply competitive. Washed cell suspensions in oxygen electrodes have been used to provide an initial comparison of the effect of iron concentration on the activity of different mineral-oxidizing bacteria (*Figure 1*; *Table 1*). The affinity constants generally appeared to reflect the growth characteristics of the bacteria as they were influenced by the concentration and oxidation state of the iron (Norris *et al.*, 1988). In batch culture, *L.ferrooxidans* was not sensitive to an Fe(I) concentration that inhibited *T.ferrooxidans*. The moderate thermophile strain ALV was the most sensitive to inhibition by Fe(III) in batch culture and, in continuous culture, was prematurely washed out before the expected critical dilution rate, possibly as a result of its relative affinities for the substrate and product iron. In chemostat culture, *L.ferrooxidans* was successful in competition for the growth-limiting Fe(II) and displaced *T.ferrooxidans* from initially mixed cultures of the bacteria. A different response of these two mesophilic bacteria to inhibitory concentrations of other base metal cations, such as copper, was used to assess the relative concentrations of each species in the mixed cultures (see later).

**The mechanism of iron oxidation**

Dugan and Lundgren (1965) envisaged a lattice of iron and sulphate several ions thick in the cell envelope of *T.ferrooxidans* which would allow the transfer of electrons to a typical cytochrome electron transport chain after the interaction of an iron−oxygen complex with an iron oxidase. The relatively direct electron transfer from the outer cell-wall through a polynuclear Fe(III) layer to the periplasm has also been proposed (Ingledew, 1982). In this scheme (*Figure 2*), the two half-reactions are separated with the generation of a trans-cytoplasmic membrane proton electrochemical potential which is utilized in ATP synthesis. Other candidates for the primary acceptor of electrons from Fe(II) include rusticyanin (Cox and Boxer, 1978), cytochrome *c* on the reducing side of rusticyanin (Blake *et al.*, 1988) and an iron−sulphur cluster, which would act with sulphate to facilitate electron transfer to an oxidized electron carrier (Fry *et al.*, 1986). Rusticyanin is a small, blue copper protein which is reducible by iron but, *in vitro* at least, at rates far too slow to account for the rate of iron oxidation in whole

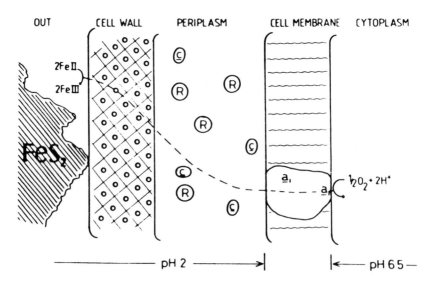

**Figure 2.** A diagrammatic representation of iron oxidation in *T.ferrooxidans* with rusticyanin, cytochrome *c* and bound iron indicated as R, *c* and the small circles in the cell wall respectively. (From Ingledew, 1986.)

102

cells (Lappin *et al.*, 1985; Blake and Shute, 1987). An iron−sulphur centre is involved in reversed electron transport in *T.ferrooxidans* (see Ingledew, 1982) but the involvement of such a centre in the process of iron oxidation requires further scrutiny. There has been a description of the simultaneous induction by iron of rusticyanin and a glycoprotein ($M_r$ 92 000) with an absorption spectrum characteristic of an iron−sulphur protein (Mjoli and Kulpa, 1988) and a report of the purification of an iron−sulphur protein ($M_r$ 63 000), which rapidly reduced cytochrome $c_{552}$ in the presence of Fe(II) (Fukumori *et al.*, 1988).

A preliminary characterization by optical spectroscopy of the various iron-oxidizing acidophiles has revealed basic differences in their respiratory chains (D.W.Barr, W.J. Ingledew and P.R.Norris, unpublished data). The difference spectra of *T.ferrooxidans* and *L.ferrooxidans* are presented as an example (*Figure 3*). The asymmetric Soret peak at 442 nm in *T.ferrooxidans* has been attributed to two cytochromes $a_1$ with alpha peak absorption at 595 nm (Cobley and Haddock, 1975) and the peaks at 419 and 551 to at least two c-type cytochromes. A high concentration of the 'downhill' electron transport chain components involved in iron oxidation has been noted, with rusticyanin and cytochrome *c* possibly each accounting for up to 5% of the cell protein (see Ingledew, 1982, for review). The difference spectrum of *L.ferrooxidans* shows major absorption peaks at 442 and 579 nm (*Figure 3*) which are attributed to a soluble, acid-stable cytochrome which has an apparent $M_r$ of slightly below 18 000. The

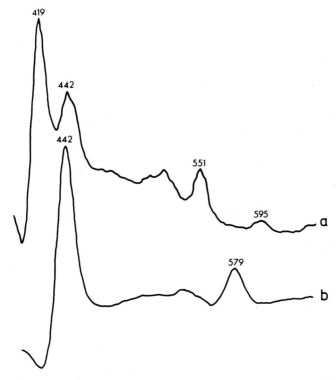

**Figure 3.** Room-temperature difference spectra (dithionite-reduced minus oxidized) of iron-grown (**a**) *T.ferrooxidans* and (**b**) *L.ferrooxidans*.

relative abundance of this novel cytochrome, a requirement for zinc during the growth of *L.ferrooxidans* on iron (see later) and the presence of zinc (as well as iron) in the cytochrome (A.Hart and P.R.Norris, unpublished data) indicate a role for it in electron transfer from iron, perhaps in a position analogous to that of rusticyanin in *T.ferrooxidans*. An absorption peak at 573 nm in iron-grown *Sulfolobus* strain BC, which was not present in the otherwise similar difference spectrum of the bacteria grown on thiosulphate, could also indicate a specific electron carrier in the iron oxidation process in this thermophilic archaebacterium (data not shown). It might perhaps be expected that the phylogenetically distinct groups of iron-oxidizing acidophiles oxidize iron by processes that differ in the precise nature of their components, as seems likely from the preliminary comparison of the respiratory chains, but which similarly convert the energy available into ATP via a chemiosmotic mechanism with the substrate oxidized in the low pH environment necessary to preclude precipitation.

## Iron reduction

The reduction of Fe(III) has been observed during the growth on sulphur of *T.thiooxidans, T.ferrooxidans* and *Sulfolobus* (Brock and Gustafson, 1976). Anaerobic growth of the sulphur-oxidizing bacteria was not demonstrated but the potential significance of the bacterial utilization of Fe(III) as an electron acceptor for sulphur oxidation under the oxygen-limited conditions prevailing in ore leach dumps was noted. The leaching of zinc from a sulphide ore by *T.ferrooxidans* anaerobically has been described (Goodman *et al.*, 1983).

Recent work on the bacterial catalysis of Fe(III) reduction with sulphur has raised several questions. An Fe(III) reducing system with sulphur as the electron donor was found to have an optimum pH of $2.8 - 3.8$ in *T.ferrooxidans* with no reduction observed below pH 1.4 or above 4.8 (Sugio *et al.*, 1985). Subsequent work with a purified sulphur $-$ Fe(III) oxidoreductase showed a pH optimum of 6.5 and no activity below pH 4.8 (Sugio *et al.*, 1987). A periplasmic location of the enzyme was suggested in an environment with a pH above 4.5 although this might be difficult to reconcile with the likelihood of Fe(III) precipitation at such a pH. A case against the necessity for iron reduction during sulphur oxidation by *T.ferrooxidans* has been presented (Corbett and Ingledew, 1987). The same rate of sulphur oxidation by oxygen and by Fe(III) and similar inhibition of the oxidation rate with both electron acceptors by HOQNO (2-*n*-heptyl-4-hydroxyquinoline *N*-oxide) indicated a similar route of electron transfer through a $bc_1$ complex rather than different systems.

## Metal toxicity

The interaction of some potentially toxic metals and mineral-oxidizing bacteria is an inevitable consequence of the solubilization of minerals in which the metals are present as major components or, with the most toxic metals, as relatively trace contaminants. The inherent toxicity of the metal cations or oxyanions and the concentrations they attain in solution can determine the extent of the contribution of bacterial activity to the dissolution of the minerals.

In commercial operations, the concentration of uranyl ions could inhibit microbial activity during leaching of uranium ores *in situ* and arsenic could prove deleterious to the bacterial oxidation of auriferous arsenopyrite concentrates in reactors. Copper concentrations up to a few g $l^{-1}$ in the dump leaching of ores might be tolerated by many acidophilic microorganisms but much higher concentrations of base metals, several tens of g $l^{-1}$, might have to be tolerated if the bacterial leaching of concentrates in reactors were to be commercialized beyond the extraction of precious metals.

The growth of some strains of *T.ferrooxidans* on Fe(II) has been reported to be inhibited by copper at $2-3$ g $l^{-1}$ (Tuovinen *et al.*, 1971) whereas others were not inhibited at 60 g $l^{-1}$ (Paknikar and Agate, 1977). Cobalt and nickel at 10 mg $l^{-1}$ inhibited the growth of *T.ferrooxidans* on thiosulphate (Tuovinen *et al.*, 1971), which contrasts with the high concentrations of base metals that are generally tolerated during growth on minerals. Copper, for example, was maintained at $30-50$ g $l^{-1}$ during the continuous bacterial leaching of a chalcopyrite concentrate (McElroy and Bruynesteyn, 1978). The toxicity of a particular metal can depend, therefore, on the bacterial strain and the growth substrate (iron, sulphur or minerals) and in some cases can also be influenced by the growth medium and the design of the toxicity assays.

The inhibition of different iron-oxidizing bacteria following the addition of metal ions to exponentially-growing cells has been described (*Figure 4* and *Table 2*; Norris *et al.*, 1986). The relatively greater sensitivity of *L.ferrooxidans* to copper is particularly striking and contrasts with a tolerance to some metals that exceeds that of *T.ferrooxidans*. The differences in the responses of these two bacteria to toxic metals and the inhibition of other acidophiles involved in mineral oxidation are considered further following a review, in order of increasing metal ion toxicity, of previous work with *T.ferrooxidans*.

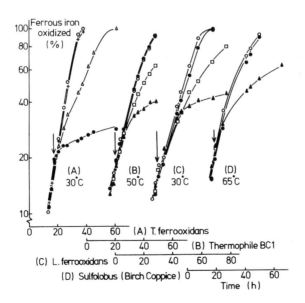

**Figure 4.** Growth-associated iron oxidation by four kinds of acidophilic bacteria with 50 mM ferrous iron and uranyl sulphate additions (↓) to give 0 (○), 0.25 (+), 0.5 (△), 1.0 (●), 2.5 (□) and 5.0 (▲) mM uranium. (From Norris *et al.*, 1986.)

**Table 2.** The concentrations of uranyl and copper sulphates, sodium molybdate, silver nitrate and mercuric chloride that gave moderate inhibition of growing, iron-oxidizing acidophilic bacteria (data derived as shown for uranium in *Figure 4*). (From Norris *et al.*, 1986.)

| Organism | Metal and concentration | | | | |
| --- | --- | --- | --- | --- | --- |
| | (mM) | | | ($\mu$M) | |
| | U | Cu | Mo | Ag | Hg |
| *T.ferrooxidans* | 0.5 | 100 | 0.25 | 0.5 | 0.25 |
| *L.ferrooxidans* | 2.5 | 1 | 1 | 5 | 0.25 |
| BC1 | 2.5 | 50 | 0.5 | 1 | 2.5 |
| *Sulfolobus* (BC) | 2.5 | 75 | 1 | 0.1 | 5 |

## *Thiobacillus ferrooxidans*

*Silver and gold.* The growth of *T.ferrooxidans* on iron has been found to be inhibited by silver at 0.1 mg $l^{-1}$ (Hoffman and Hendrix, 1976) and the lag phase before growth was extended by only 0.1 $\mu$g $l^{-1}$ or 1 nM silver (Norris and Kelly, 1978). The inhibition of exponentially-growing bacteria by 0.5 $\mu$M silver (*Table 2*) was followed by a resumption of growth at the control rate after four days further incubation (data not shown). Adaptation to greater tolerance via serial culture in medium containing increasing concentrations of silver produced bacteria which were relatively silver-resistant, apparently by virtue of a reduced capacity for accumulation of the metal (Norris and Kelly, 1978). Growth was eventually obtained in a medium to which 0.1 mM $AgNO_3$ has been added although some precipitation was observed in this case. In a similar fashion, a silver-resistant culture, which grew in the presence of 0.5 mM $AgNO_3$, was obtained (Sugio *et al.*, 1981); the cytochrome oxidase was suggested as a possible site of inhibition by silver but resistance was again attributed to reduced penetration of the metal across the cell membrane. The progressive accumulation of silver by the cells could entail some binding to exposed sites following the disruption of the membrane permeability barrier, which was indicated by a rapid and massive loss of cell potassium (Norris and Kelly, 1978), rather than by genuine ion transport. The uptake of silver was not affected by the presence of a 10-fold excess of copper (Norris and Kelly, 1978) in contrast to competition between these metals for common sites of action or transport in *Escherichia coli* (Ghandour *et al.*, 1988).

Although *T.ferrooxidans* could not grow on silver sulphide, the presence of the mineral did not affect growth on sulphur or other sulphides (Norris and Kelly, 1978). Furthermore, silver in other mineral sulphides did not affect the bacterial activity but resulted in the deposition of silver sulphide on the bacterial surface up to a concentration of 25% of the bacterial dry weight (Pooley, 1982). The formation of silver sulphides would also preclude the presence of sufficient silver ions in solution to inhibit the bacteria when silver is used, up to 400 mg $l^{-1}$ for example (Bruynesteyn *et al.*, 1983), to stimulate the bacterial leaching of chalcopyrite. The catalytic action of the silver is probably through argentite ($Ag_2S$) formation at the mineral surface which disrupts an otherwise less porous layer of elemental sulphur which can reduce the rate and extent of mineral oxidation by Fe(III) (Miller and Portillo, 1981).

*T.ferrooxidans* has been used successfully in the extraction of gold from refractory sulphide concentrates, and there are no reports of the bacterial activity in the process being inhibited by the gold. Some slight inhibition of growth on iron by trace concentrations of chloroaurate has been noted (Norris and Kelly, 1978) but, in contrast to the quantitatively similar inhibition by silver, it was progressively reduced with an increase in concentration so that 0.1 mM sodium chloroaurate, which is likely to be transformed in the acidic, iron-containing medium, produced a shorter lag phase than in control cultures. A similar reduction of the lag phase by equimolar sodium chloride (Norris and Kelly, 1978) and the apparent inhibition of *T.ferrooxidans* by potassium and sodium at trace levels but not at higher concentrations (Tuovinen and Kelly, 1974a) indicate the limited significance of such demonstrations of the complex metal ion toxicity. The adsorption of gold to the cell-wall and membrane of *T.ferrooxidans* has been described with some indication of the reduction of ionic gold to the metal (Pivovarova *et al.*, 1986).

*Mercury.* The initiation of growth of *T.ferrooxidans* on iron was found to be inhibited by 1 $\mu$M $HgCl_2$ but growth did occur after an extended lag phase (Norris and Kelly, 1978). The inhibition of exponentially-growing bacteria by 0.25 $\mu$M (*Table 2*) and even 1 $\mu$M $HgCl_2$ was transient whereas 2 $\mu$M $HgCl_2$ prevented further growth. The reduction and volatilization of mercury by *T.ferrooxidans* has been described with strains that were resistant to, or adapted to grow in the presence of, 5 $\mu$M (Olson *et al.*, 1982) and 10 $\mu$M $HgCl_2$ (Booth and Williams, 1984).

Intact *T.ferrooxidans* volatilized mercury at pH 2.5 whereas the optimum pH for the activity of cell-free extracts (Olson *et al.*, 1982) and of the purified mercuric reductase (Booth and Williams, 1984) was about neutrality as would be expected of an intracellular enzyme. Antisera against purified mercuric reductases, which inactivated the enzymes from a range of Gram-negative bacteria, did not inactivate the enzyme from *T.ferrooxidans*; the latter enzyme also showed some different sensitivities from that of *E.coli* to some other metals but it was otherwise found to be structurally and functionally similar to those from other Gram-negative bacteria. $K_m$ values of 15 $\mu$M $HgCl_2$ for activity of crude cell extracts (Olson *et al.*, 1982) and 9 $\mu$M with the purified enzyme (Booth and Williams, 1984) were similar to values found with *E.coli* and *Pseudomonas aeruginosa* under similar conditions. Resistance to mercury among Gram-negative and Gram-positive bacteria is frequently inducible and plasmid-encoded but appears to be constitutive and chromosomally-determined in *T.ferrooxidans*.

The possible significance of mercury resistance to the activity of *T.ferrooxidans* in the mobilization of mercury in acidic, mineral-rich environments or to its activity in the leaching of mercury-containing ores has been indicated by the demonstration of the oxidation of pyrite in pyrite−cinnabar (HgS) mixtures by a mercury-resistant strain in the absence of any activity of mercury-sensitive bacteria (Baldi and Olson, 1987). Elemental mercury was produced only by the resistant strain from the mineral mixtures and from cinnabar alone although the latter could not support growth.

*Uranium and thorium.* The growth on Fe(II) of *T.ferrooxidans* strains that have not been exposed to uranium previously is generally inhibited by 0.5−1 mM uranyl ions (Tuovinen and Kelly, 1974b). Similar concentrations of uranium in solution have been recorded at a bacterial leaching operation (see Tuovinen *et al.*, 1981) but *T.ferrooxidans* can readily develop tolerance to even 5 mM $UO_2^{2+}$ (Tuovinen and Kelly,

1974b; DiSpirito and Tuovinen, 1982). In contrast to only 10% inhibition of iron oxidation by 5 mM $UO_2^{2+}$, the fixation of carbon dioxide was 96% and 68% inhibited in unadapted and tolerant (adapted) bacteria, respectively (Tuovinen and Kelly, 1974c). As noted earlier with the toxicity and binding of silver, this could indicate less membrane damage and a decreased permeability to the toxic metal in adapted bacteria rather than a reduced sensitivity of a particular aspect of metabolism (carbon dioxide fixation in this case) as a basis for resistance. Significant accumulation of uranium could not be demonstrated with growing bacteria (Tuovinen and Kelly, 1974b,c) but accumulation by washed cell suspensions was approximately proportional to the $UO_2^{2+}$ concentration with up to 63 $\mu$g U mg protein$^{-1}$ accumulated from 10 mM $UO_2^{2+}$ (DiSpirito *et al.*, 1983). The accumulation of uranium reached equilibrium within an hour, was not affected by concurrent iron oxidation or by equimolar concentrations of several other metals and was mostly associated with the cell-wall and membrane.

The simultaneous bacterial leaching of uranium and thorium has been described, the concentration of the latter being the greater in some leach liquors (Tuovinen *et al.*, 1981). Various isolates of *T.ferrooxidans* were equally or only slightly less sensitive to $Th^{4+}$ than to $UO_2^{2+}$ (DiSpirito and Tuovinen, 1982).

*Thallium.* The growth of *T.ferrooxidans* on iron can be inhibited by greater than 0.1 mM thallium in a medium containing only trace levels of potassium (Tuovinen and Kelly, 1974a). In contrast to the other metal ions, silver, mercury and uranium, which can be particularly toxic to *T.ferrooxidans*, the uptake of $Tl^+$ by bacteria and yeast has been readily associated with the transport of an essential element, by virtue of the same charge and similar radii of $K^+$ and $Tl^+$ ions (Norris *et al.*, 1976). The significant alleviation by $K^+$ of $Tl^+$ and $Rb^+$ toxicity to *T.ferrooxidans* (Tuovinen and Kelly, 1974a) could therefore be attributed to competition in ion transport and contrasts with the alleviation of uranium toxicity relatively non-specifically by greater than 50 mM $K^+$ or other monovalent ions (Tuovinen and Kelly, 1974a).

*Copper, cobalt, nickel, zinc and cadmium.* The relative toxicity of these divalent metal cations to *T.ferrooxidans* appears to vary with the strain and growth conditions. For example, copper has been found to be more toxic (Tuovinen *et al.*, 1971; P.R.Norris, unpublished work with *T.ferrooxidans* DSM 583) and less toxic than cobalt (Barbic, 1977; Paknikar and Agate, 1988). Moderate inhibition of growing bacteria was caused by 100 mM copper (*Table 2*) but even near complete inhibition by 250 mM copper was reversible, rapid growth resuming after three days further incubation. In such cases, and where adaptation through serial culture involves relatively low concentrations of these divalent metal cations, it is probable that all of the affected cells become adapted in contrast to the selection of resistant mutants following exposure to higher and relatively more toxic concentrations (Groudeva *et al.*, 1981).

The tolerance to high concentrations of these metals by *T.ferrooxidans* can be attributed to the limited binding or accumulation of the metals by the bacteria. Negligible accumulation of copper, zinc and cadmium was found in comparison with that of uranium by suspensions of non-growing cells (DiSpirito *et al.*, 1983), although some accumulation of nickel and cobalt, tolerance to which would also probably require exclusion from sensitive sites, necessitates some caution in relating the observations to the interactions

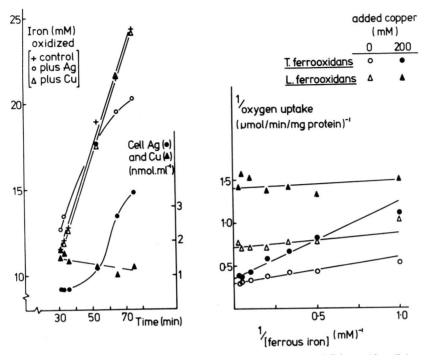

**Figure 5.** (Left) Iron oxidation and metal accumulation by cell suspensions of *T.ferrooxidans* (0.1 mg dry weight ml$^{-1}$) in acidified (H$_2$SO$_4$) water at pH 1.5 and 30°C. Additions of AgNO$_3$ and CuSO$_4$ were made to give 10 $\mu$M metal concentrations in separate suspensions 30 min after the addition of ferrous sulphate (50 mM). Samples were washed on filters with water (pH 1.5) as in similar, earlier experiments. (Norris and Kelly, 1978.)

**Figure 6.** (Right) Reciprocal plots of oxygen uptake by cell suspensions of mesophilic, iron-oxidizing bacteria at pH 1.7, 30°C and the indicated concentrations of ferrous and copper sulphates. (From Norris *et al.*, 1988.)

that might occur with the metals during growth of the bacteria. In contrast to the accumulation of silver and the consequent inhibition of iron oxidation noted earlier, some association of copper with cells possibly decreased with time (*Figure 5*); although the location of the copper and whether it was really bound to specific sites was not investigated, the lack of inhibition of iron oxidation or any loss of cell potassium indicated the preservation of membrane integrity. In the apparent absence of copper uptake, it is possible that, where the growth of *T.ferrooxidans* continues in the presence of high concentrations of the metal, though at a reduced rate, the metal could be exerting an effect on the oxidation of iron at the cell surface. *Figure 6* illustrates the essentially competitive nature of inhibition by copper of iron oxidation.

## Leptospirillum ferrooxidans

The greater sensitivity of *L.ferrooxidans* than of *T.ferrooxidans* to copper and some other divalent metal cations has been observed with the inhibition of iron oxidation by non-growing cell suspensions (Eccleston *et al.*, 1985) and, at much lower metal concentrations, with the inhibition of growth (Norris *et al.*, 1986; *Table 2*).

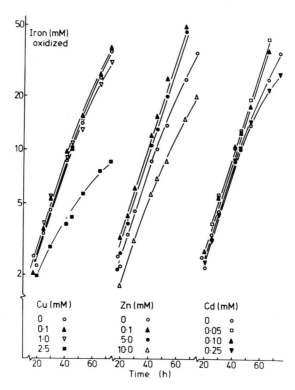

**Figure 7.** Iron oxidation during growth of *L.ferrooxidans* at 30°C in medium (pH 1.7) containing the indicated concentrations of metal sulphates.

The uncompetitive inhibition by copper of iron oxidation by *L.ferrooxidans* (Norris *et al.*, 1988; *Figure 6*) in contrast to the predominantly competitive inhibition with *T.ferrooxidans* means that the relative inhibition of the activity of the cell suspensions depends on the concentration of the substrate Fe(II). This difference was the basis of the selective inhibition of *L.ferrooxidans* by copper in samples from mixed cultures when 50 mM Fe(II) was added to the samples from chemostats in an oxygen electrode to give an estimate of relative concentrations of the two bacteria during competition experiments (Norris *et al.*, 1988). Cadmium also competitively and uncompetitively inhibited iron oxidation by cell suspensions of *T.ferrooxidans* and *L.ferrooxidans* respectively, perhaps suggesting an influence on the access of $Fe^{2+}$ to the oxidation sites which was specific for the different bacteria but not for a particular divalent metal cation.

A further examination of the effects of some of the divalent metal cations on the growth of *L.ferrooxidans* revealed the requirement for zinc noted earlier in discussion of the respiratory chains involved in iron oxidation. Copper was slightly less inhibitory (*Figure 7*) than indicated previously when it was added to growing cells (*Table 2*), and other isolates of *Leptospirillum* were inhibited only at between 5 and 10 mM copper (Norris *et al.*, 1988). Whatever the strain or conditions, the greater sensitivity in comparison with *T.ferrooxidans* remains clear. Some inhibition of growth by 10 mM zinc was evident but concentrations below 5 mM resulted in the maintenance of a higher

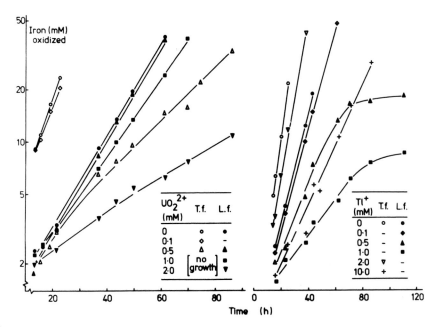

**Figure 8.** The effects of uranyl sulphate and thallium nitrate on growth-associated iron oxidation by *T.ferrooxidans* (T.f.) and *L.ferrooxidans* (L.f.) at pH 1.7 and 30°C in a medium containing 2.5 mM K$^+$.

rate of growth-associated iron oxidation than in control cultures (*Figure 7*). A slightly higher rate of iron oxidation was maintained in medium to which 0.1 µM zinc was added while the maximum effect was obtained at 5 µM zinc with higher concentrations being superfluous. Cadmium, but not copper, nickel, cobalt or manganese, could also be beneficial but 50 times more cadmium than zinc was required to produce a quantitatively similar effect and the onset of inhibition occurred with about 20 times less of the more toxic metal (*Figure 7*).

*L.ferrooxidans* was less sensitive than *T.ferrooxidans* to silver, molybdenum and uranium when the metals were added to growing bacteria (*Table 2*) and following inoculation of medium containing the metals, as shown for uranium (*Figure 8*). The relative susceptibility of the bacteria was reversed with thallium. The growth rate of *T.ferrooxidans* was considerably reduced by 10 mM thallium nitrate but sodium nitrate was also inhibitory at this concentration, reflecting the well-documented sensitivity to some anions (for example, Barbic, 1977; Ingledew, 1982). The inhibition of *L.ferrooxidans*, which is less sensitive to nitrate (Norris *et al.*, 1988), could have resulted from some progressive accumulation of thallium via the potassium transport pathway until a bactericidal, intracellular concentration was reached. There has been no study of whether the greater sensitivity of *L.ferrooxidans* reflects less discrimination between thallium and potassium during the transport process or a greater susceptibility to the metal once accumulated.

The toxicity of arsenic, which is likely to depend on its oxidation state and complex formation, is of some concern with the increasing interest in the bacterial pretreatment of auriferous arsenopyrite concentrates to facilitate the recovery of the gold. A reduced rate of growth of *L.ferrooxidans* has been noted in the presence of 5 mM sodium arsenate while with 10 mM arsenate the growth of *T.ferrooxidans* was not affected but some iron−arsenic complex formation and precipitation occurred (Norris *et al.*, 1988).

## Thermophilic bacteria

There have been very few studies of the interactions of the moderately thermophilic iron-oxidizing bacteria with toxic metals. The growth of strain BC1 was inhibited by concentrations of metals roughly similar to those that inhibited *T.ferrooxidans* although slightly less by uranium and more by copper (*Table 2*). Some differences in the sensitivity of strains to metals have been observed both in the inhibition of respiration by cell suspensions (Brierley and Brierley, 1986) and in the inhibition of growth (P.R.Norris, unpublished data). Strain ALV and strain TH3 were more and less sensitive, respectively, than strain BC1 to inhibition by copper and strain TH3 was also more resistant than strain TH1 to uranium.

Strain BC1 and *Sulfolobus* strain BC were less sensitive than the mesophiles to mercury (*Table 2*). However, reaction with thiosulphate (or its decomposition products), which was present in the medium as a source of reduced sulphur and which is required by these bacteria during autotrophic growth on iron (Norris and Barr, 1985; Marsh *et al.*, 1983), could have reduced the metal ion concentration. The anaerobic reduction of molybdate during the growth of *Sulfolobus* species on sulphur has been described and the bacteria appeared markedly more resistant than *T.ferrooxidans* to the metal ion with growth only slightly inhibited at 7.8 mM molybdate (Brierley, 1974; Brierley and Brierley, 1982). Mineral-oxidizing strains of *Sulfolobus* have been adapted to increasing concentrations of nickel and copper and some growth and oxidation of chalcopyrite in the presence of copper at 30 g $1^{-1}$ has been noted although the bacteria were more active at lower copper concentrations (Norris and Parrott, 1986).

## Thiobacillus thiooxidans

*T.thiooxidans* can grow on sulphur or sulphide released from minerals following oxidation of their metal moiety by *L.ferrooxidans* or *T.ferrooxidans* (Norris, 1983) and can degrade some minerals in the absence of the iron-oxidizing bacteria (Khalid and Ralph, 1977; Lizama and Suzuki, 1988). The bacteria also have a role in biological oxidation ponds for the treatment of ore-mill process waters containing thiosalts. In this last connection, a *T.thiooxidans* strain was generally more sensitive to metals than a mixed culture from the outlet of a pond (Silver and Dinardo, 1981). The growth of the mixed culture on thiosulphate was inhibited at concentrations of zinc and cadmium above 200 and 300 mg $1^{-1}$, respectively, but was 75% inhibited by copper at only 5 mg $1^{-1}$. The adaptation of several strains of *T.thiooxidans* to tolerance of a cobalt concentration of 25 g $1^{-1}$ (Groudev, 1981) has indicated that, as noted earlier with reference to *T.ferrooxidans*, the growth on elemental sulphur is less sensitive to inhibition by divalent metal cations than growth on the thiosalts. Molybdate and chromate prevented the growth of *T.thiooxidans* on sulphur in the range 0.2−0.5 mM while the vanadate ion was reduced to the less toxic vanadyl ion which was not toxic at 50 mM (Jack *et al.*, 1980).

**Discussion**

This chapter has illustrated some differences in the behaviour of the various iron-oxidizing bacteria towards metal ions as substrates and inhibitors. Many of the differences, such as in the kinetics of iron oxidation, are of degree in response to challenge by the metals and their significance in the context of mineral oxidation is uncertain. For example, the capacity of *L.ferrooxidans* to compete successfully with *T.ferrooxidans* and even become dominant during the growth on pyrite of initially mixed cultures could partly be derived from its higher affinity for Fe(II) and greater tolerance of Fe(III). The concentration of the Fe(III) increases during pyrite oxidation but the increasing acidity, also better tolerated by *L.ferrooxidans*, could also be a significant factor in determining different limits to favourable growth conditions for these bacteria (Norris, 1988). The growth of *Sulfolobus* on pyrite at high temperatures, unlike the growth of the mesophiles at lower temperatures, has been shown to proceed with a significant concentration of Fe(II) in solution (Norris and Parrott, 1986). Therefore, although the ratios of $K_m$ to $K_i$ values for Fe(II) and Fe(III), respectively, were similar with *Sulfolobus* and *T.ferrooxidans* (*Table 1*), the potential for competitive inhibition of iron oxidation by the end product during growth of the thermophile could be less than during growth of the mesophile.

The illustrated differences in sensitivity to some metals, particularly between *L.ferrooxidans* and *T.ferrooxidans*, should also be seen in a wider perspective. *L.ferrooxidans* was more tolerant than *T.ferrooxidans* to uranium but the capacity of the latter to adapt to higher concentrations of this and other metals is well established while there has not been a similar assessment of the adaptability of *L.ferrooxidans*. The growth of *L.ferrooxidans* on chalcopyrite in the presence of copper at 25 g $l^{-1}$ has been noted however (Norris *et al.*, 1988). This contrasts notably with the inhibition of its growth on Fe(II) by copper concentrations that were far lower, and particularly so in comparison to the concentrations required to inhibit the other iron-oxidizing bacteria. In summary, mercury and silver ions appear lethal to all of the bacteria at sub-millimolar concentrations with the cell membrane a likely site of action. Uranium is inhibitory at higher concentrations but still considerably more toxic than a range of divalent metal cations except where these inhibit *L.ferrooxidans*. The sensitivity of the latter to these divalent metal cations could reflect their uptake during growth in contrast to their exclusion by *T.ferrooxidans*, as suggested to account for the different sensitivity of these bacteria to thallium, but there is presently no evidence to support this possibility. As the cell−metal interactions with bacteria other than *T.ferrooxidans* are relatively poorly studied, the extent of the differences in the iron oxidation mechanisms is also currently unknown.

A review of the progress in the development of molecular genetic work with *T.ferrooxidans* (Woods *et al.*, 1986) has noted that metal ion tolerance, with its obvious industrial relevance in these bacteria, would be an attractive marker for genetic studies. However, the evidence for plasmid-borne resistance to uranium (Martin *et al.*, 1983), mercury and silver (Visca *et al.*, 1986) in *T.ferrooxidans* remains circumstantial. The isolation of the rusticyanin gene (Jedlicki *et al.*, 1986; Kulpa *et al.*, 1986) heralds the application of molecular biology techniques in the study of the iron oxidation process. The demonstration of phenotypic variation in *T.ferrooxidans*, which includes the reversible loss of iron oxidation and is apparently caused by the mobility of repeated DNA sequences (Holmes *et al.*, 1988), should also promote some advantageous integration of genetic and biochemical studies in the investigation of the cell−metal interactions.

## References

Baldi,F. and Olson,G.J. (1987) Effects of cinnabar on pyrite oxidation by *Thiobacillus ferrooxidans* and cinnabar mobilization by a mercury-resistant strain. *Applied and Environmental Microbiology* **53**, 772−776.

Balashova,V.V., Vedinina,I.Ya., Markosyan,G.E. and Zavarzin,G.A. (1974) The autotrophic growth of *Leptospirillum ferrooxidans. Microbiology* **43**, 491−494.

Barbic,F.F. (1977) Effects of different compounds of metals and of their mixtures on the growth and survival of *Thiobacillus ferrooxidans. Zeitschrift für Allgemeine Mikrobiologie* **17**, 277−281.

Blake,R.C. and Shute,E.A. (1987) Respiratory enzymes of *Thiobacillus ferrooxidans*. A kinetic study of electron transport between iron and rusticyanin in sulfate media. *Journal of Biological Chemistry* **262**, 14983−14989.

Blake,R.C., White,K.J. and Shute,E.A. (1988) Electron transfer from Fe(II) to rusticyanin is catalyzed by an acid-stable cytochrome. In *Biohydrometallurgy (Proceedings of the International Symposium, Warwick 1987)* (eds. P.R.Norris and D.P.Kelly), Science and Technology Letters, Kew, pp. 103−110.

Booth,J.E. and Williams,J.W. (1984) The isolation of a mercuric ion-reducing flavoprotein from *Thiobacillus ferrooxidans. Journal of General Microbiology* **130**, 725−730.

Brierley,C.L. (1974) Molybdenite leaching: use of a high-temperature microbe. *Journal of the Less Common Metals* **36**, 237−247.

Brierley,C.L. and Brierley,J.A. (1982) Anaerobic reduction of molybdenum by *Sulfolobus* species. *Zentralblatt für Bakteriologie, Mikrobiologie und Hygiene. I. Abt. Originale, Series* C **3**, 289−294.

Brierley,J.A. and Brierley,C.L. (1986) Microbial mining using thermophilic microorganisms. In *Thermophiles: General, Molecular and Applied Microbiology* (ed. T.D.Brock), Wiley, New York, pp. 279−305.

Brock,T.D. and Gustafson,J. (1976) Ferric iron reduction by sulfur- and iron-oxidizing bacteria. *Applied and Environmental Microbiology* **32**, 567−571.

Bruynesteyn,A., Lawrence,R.W., Vizsolyi,A. and Hackl,R. (1983) An elemental sulphur-producing biohydrometallurgical process for treating sulphide concentrates. In *Recent Progress in Biohydrometallurgy* (eds. G.Rossi and A.E.Torma), Associazione Mineraria Sarda, Iglesias, pp. 151−168.

Cobley,J.G. and Haddock,B.A. (1975) The respiratory chain of *Thiobacillus ferrooxidans*: the reduction of cytochromes by $Fe^{2+}$ and the preliminary characterization of rusticyanin, a novel blue copper protein. *FEBS Letters* **60**, 29−33.

Corbett,C.M. and Ingledew,W.J. (1987) Is $Fe^{3+/2+}$ cycling an intermediate in sulphur oxidation by $Fe^{2+}$-grown *Thiobacillus ferrooxidans*? *FEMS Microbiology Letters* **41**, 1−6.

Cox,J.C. and Boxer,D.H. (1978) The purification and some properties of rusticyanin, a blue copper protein involved in iron(II) oxidation from *Thiobacillus ferrooxidans. Biochemical Journal* **174**, 497−502.

DiSpirito,A.A. and Tuovinen,O.H. (1982) Kinetics of uranous ion and ferrous iron oxidation by *Thiobacillus ferrooxidans. Archives of Microbiology* **133**, 33−37.

DiSpirito,A.A., Talnagi,J.W. and Tuovinen,O.H. (1983) Accumulation and cellular distribution of uranium in *Thiobacillus ferrooxidans. Archives of Microbiology* **135**, 250−253.

Dugan,P.R. and Lundgren,D.G. (1965) Energy supply for the chemoautotroph *Ferrobacillus ferrooxidans. Journal of Bacteriology* **89**, 825−834.

Eccleston,M., Kelly,D.P. and Wood,A.P. (1985) Autotrophic growth and iron oxidation and inhibition kinetics of *Leptospirillum ferrooxidans*. In *Planetary Ecology* (eds. D.E.Caldwell, J.A.Brierley and C.L.Brierley), Van Nostrand Reinhold, New York, pp. 263−272.

Fry,I.V., Lazaroff,N. and Packer,L. (1986) Sulfate-dependent iron oxidation by *Thiobacillus ferrooxidans*: characterization of a new EPR detectable electron transport component on the reducing side of rusticyanin. *Archives of Biochemistry and Biophysics* **246**, 650−654.

Fukumori,Y., Yano,T., Sato,A. and Yamanaka,T. (1988) Fe(II)-oxidizing enzyme purified from *Thiobacillus ferrooxidans. FEMS Microbiology Letters* **50**, 169−172.

Ghandour,W., Hubbard,J.A., Deistung,J., Hughes,M.N. and Poole,R.K. (1988) The uptake of silver ions by *Escherichia coli* K12: toxic effects and interaction with copper ions. *Applied and Environmental Microbiology* **28**, 559−565.

Goodman,A.E., Babij,T. and Ritchie,A.I.M. (1983) Leaching of a sulphide ore by *Thiobacillus ferrooxidans* under anaerobic conditions. In *Recent Progress in Biohydrometallurgy* (eds. G.Rossi and A.E.Torma), Associazione Mineraria Sarda, Iglesias, pp. 361−376.

Groudev,S.N. (1981) Leaching of cobalt from synthetic cobalt sulphide by *Thiobacillus ferrooxidans* and *Thiobacillus thiooxidans*. *Comptes Rendus de l'Académie bulgare des Sciences* **34**, 217−220.

Groudeva,V.I., Groudev,V.I. and Markov,K.I. (1981) Selection of *Thiobacillus ferrooxidans* mutants tolerant to high concentrations of copper ions. *Comptes Rendus de l'Académie bulgare des Sciences* **34**, 375−378.

Harrison,A.P. (1986) Characteristics of *Thiobacillus ferrooxidans* and other iron-oxidizing bacteria, with emphasis on nucleic acid analyses. *Biotechnology and Applied Biochemistry* **8**, 249−257.

Hoffman,L.E. and Hendrix,J.L. (1976) Inhibition of *Thiobacillus ferrooxidans* by soluble silver. *Biotechnology and Bioengineering* **18**, 1161−1165.

Holmes,D.S., Yates,J.R. and Schrader,J. (1988) Mobile, repeated DNA sequences in *Thiobacillus ferrooxidans* and their significance for biomining. In *Biohydrometallurgy (Proceedings of the International Symposium, Warwick 1987)* (eds, P.R.Norris and D.P.Kelly), Science and Technology Letters, Kew, pp. 153−160.

Hutchins,S.R., Davidson,M.S., Brierley,J.A. and Brierley,C.L. (1986) Microorganisms in reclamation of metals. *Annual Review of Microbiology* **40**, 311−336.

Ingledew,W.J. (1982) *Thiobacillus ferrooxidans*: the bioenergetics of an acidophilic chemolithotroph. *Biochimica et Biophysica Acta* **683**, 89−117.

Ingledew,W.J. (1986) Ferrous iron oxidation by *Thiobacillus ferrooxidans*. In *Workshop on Biotechnology for the Mining Metal-Refining and Fossil Fuel Processing Industries* (eds. H.L.Ehrlich and D.S.Holmes), Wiley, New York, pp. 23−33.

Jack,T.R., Sullivan,E.A. and Zajic,J.E. (1980) Growth inhibition of *Thiobacillus thiooxidans* by metals and reductive detoxification of vanadium(V). *European Journal of Applied Microbiology and Biotechnology* **9**, 21−30.

Jedlicki,E., Reyes,R., Jordana,X., Mercereau-Puijalon,O. and Allende,J.E. (1986) Rusticyanin: initial studies on the regulation of its synthesis and gene isolation. *Biotechnology and Applied Biochemistry* **8**, 342−350.

Jones,C.A. and Kelly,D.P. (1983) Growth of *Thiobacillus ferrooxidans* on ferrous iron in chemostat culture: Influence of product and substrate inhibition. *Journal of Chemical Technology and Biotechnology* **33B**, 241−261.

Karavaiko,G.I., Golovacheva,R.S., Pivovarova,T.A., Tzaplina,I.A. and Vartanjan,N.S. (1988) Thermophilic bacteria of the genus *Sulfobacillus*. In *Biohydrometallurgy (Proceedings of the International Symposium, Warwick 1987)* (eds. P.R.Norris and D.P.Kelly), Science and Technology Letters, Kew, pp. 29−41.

Kelly,D.P. and Jones,C.A. (1978) Factors affecting metabolism and ferrous iron oxidation in suspensions and batch cultures of *Thiobacillus ferrooxidans*: relevance to ferric iron leach solution regeneration. In *Metallurgical Applications of Bacterial Leaching and Related Microbiological Phenomena* (eds. L.E.Murr, A.E.Torma and J.A.Brierley), Academic Press, New York, pp. 19−44.

Kelly,D.P., Eccleston,M. and Jones,C.A. (1977) Evaluation of continuous cultivation of *Thiobacillus ferrooxidans* on ferrous iron or tetrathionate. In *Conference: Bacterial Leaching* (ed. W.Schwartz), Verlag Chemie, Weinheim, pp. 1−7.

Kelly,D.P., Norris,P.R. and Brierley,C.L. (1979) Microbiological methods for the extraction and recovery of metals. In *Microbial Technology: Current State, Future Prospects* (eds. A.T.Bull, C.Ratledge and D.C.Ellwood), Cambridge University Press, Cambridge, pp. 263−308.

Khalid,A.M. and Ralph,B.J. (1977) The leaching behaviour of various zinc sulphide minerals with three *Thiobacillus* species. In *Conference: Bacterial Leaching* (ed. W.Schwartz), Verlag Chemie, Weinheim, pp. 261−270.

Kulpa,C.F., Roskey,M.T. and Mjoli,N. (1986) Construction of genomic libraries and induction of iron oxidation in *Thiobacillus ferrooxidans*. *Biotechnology and Applied Biochemistry* **8**, 330−341.

Lappin,A.G., Lewis,C.A. and Ingledew,W.J. (1985) Kinetics and mechanisms of reduction of rusticyanin, a blue copper protein from *Thiobacillus ferrooxidans*, by inorganic cations. *Inorganic Chemistry* **24**, 1446−1450.

Lewis,A.J. and Miller,J.D.A. (1977) Stannous and cuprous ion oxidation by *Thiobacillus ferrooxidans*. *Canadian Journal of Microbiology* **23**, 319−324.

Lizama,H.M. and Suzuki,I.(1988) Bacterial leaching of a sulfide ore by *Thiobacillus ferrooxidans* and *Thiobacillus thiooxidans*: 1. Shake flask studies. *Biotechnology and Bioengineering* **32**, 110−116.

Marsh,R.M. and Norris,P.R. (1983) The isolation of some thermophilic, autotrophic, iron- and sulphur-oxidizing bacteria. *FEMS Microbiology Letters* **17**, 311−315.

Marsh,R.M., Norris,P.R. and Le Roux,N.W. (1983) Growth and mineral oxidation studies with *Sulfolobus*. In *Recent Progress in Biohydrometallurgy* (eds. G.Rossi and A.E.Torma), Associazione Mineraria Sarda, Iglesias, pp. 71–81.

Martin,P.A.W., Dugan,P.R. and Tuovinen,O.H. (1983) Uranium resistance of *Thiobacillus ferrooxidans*. *European Journal of Applied Microbiology and Biotechnology* **18**, 392–395.

McElroy,R.O. and Bruynesteyn,A. (1978) Continuous biological leaching of chalcopyrite concentrates: demonstration and economic analysis. In *Metallurgical Applications of Bacterial Leaching and Related Microbiological Phenomena* (eds. L.E.Murr, A.E.Torma and J.A.Brierley), Academic Press, New York, pp. 441–462.

Miller,J.D. and Portillo,H.Q. (1981) Silver catalysis in ferric sulphate leaching of chalcopyrite. *Developments in Mineral Processing* **2**, 851–901.

Mjoli,N. and Kulpa,C.F. (1988) Identification of a unique outer membrane protein required for iron oxidation in *Thiobacillus ferrooxidans*. In *Biohydrometallurgy (Proceedings of the International Symposium, Warwick 1987)* (eds. P.R.Norris and D.P.Kelly), Science and Technology Letters, Kew, pp. 89–102.

Norris,P.R. (1983) Iron and mineral oxidation with *Leptospirillum*-like bacteria. In *Recent Progress in Biohydrometallurgy* (eds. G.Rossi and A.E.Torma), Associazione Mineraria Sarda, Iglesias, pp. 83–96.

Norris,P.R. (1988) Bacterial diversity in reactor mineral leaching. In *Proceedings of the 8th International Biotechnology Symposium, Paris, 1988* (in press).

Norris,P.R. and Barr,D.W. (1985) Growth and iron oxidation by acidophilic moderate thermophiles. *FEMS Microbiology Letters* **28**, 221–224.

Norris,P.R. and Kelly,D.P. (1978) Toxic metals in leaching systems. In *Metallurgical Applications of Bacterial Leaching and Related Microbiological Phenomena* (eds. L.E.Murr, A.E.Torma and J.A.Brierley), Academic Press, New York, pp. 83–102.

Norris,P.R. and Parrott,L. (1986) High temperature, mineral concentrate dissolution with *Sulfolobus*. In *Fundamental and Applied Biohydrometallurgy*. (eds. R.W.Lawrence, R.M.R.Branion and H.G.Ebner), Elsevier, Amsterdam, pp. 355–365.

Norris,P.R., Man,W.K., Hughes,M.N. and Kelly,D.P. (1976) Toxicity and accumulation of thallium in bacteria and yeast. *Archives of Microbiology* **110**, 279–286.

Norris,P.R., Parrott,L. and Marsh,R.M. (1986) Moderately thermophilic mineral-oxidizing bacteria. In *Workshop on Biotechnology for the Mining, Metal-Refining and Fossil Fuel Processing Industries* (eds. H.L.Ehrlich and D.S.Holmes), Wiley, New York, pp. 253–262.

Norris,P.R., Barr,D.W. and Hinson,D. (1988) Iron and mineral oxidation by acidophilic bacteria: affinities for iron and attachment to pyrite. In *Biohydrometallurgy (Proceedings of the International Symposium, Warwick 1987)* (eds. P.R.Norris and D.P.Kelly), Science and Technology Letters, Kew, pp. 43–59.

Olson,G.J., Porter,F.D., Rubenstein,J. and Silver,S. (1982) Mercuric reductase enzyme from a mercury-volatilizing strain of *Thiobacillus ferrooxidans*. *Journal of Bacteriology* **151**, 1230–1236.

Paknikar,K.M. and Agate,A.D. (1988) Occurrence of a *Thiobacillus ferrooxidans* strain tolerating unusually high concentrations of metals and an associated metal-tolerant acidophilic heterotrophic bacterium. In *Biohydrometallurgy (Proceedings of the International Symposium, Warwick 1987)* (eds. P.R.Norris and D.P.Kelly), Science and Technology Letters, Kew, pp. 558–560.

Pivovarova,T.A. and Golovacheva,R.S. (1985) Microorganisms important for hydrometallurgy: cytology, physiology and biochemistry. In *Biogeotechnology of Metals* (eds. G.I.Karavaiko and S.N.Groudev), UNEP, Centre of International Projects, GKNT, Moscow, pp. 27–55.

Pivovarova,T.A., Markosyan,G.E. and Karavaiko,G.I. (1981) Morphogenesis and fine structure of *Leptospirillum ferrooxidans*. *Microbiology* **50**, 339–344.

Pivovarova,T.A., Korobushkina,E.D., Krasheninnikova,S.A., Rubtsov,A.E. and Karavaiko,G.I. (1986) Influence of gold ions on *Thiobacillus ferrooxidans*. *Microbiology* **55**, 774–780.

Pooley,F.D. (1982) Bacteria accumulate silver during leaching of sulphide ore minerals. *Nature (London)* **296**, 642–643.

Silver,M. and Dinardo,O. (1981) Factors affecting oxidation of thiosalts by thiobacilli. *Applied and Environmental Microbiology* **41**, 1301–1309.

Sugio,T., Tano,T. and Imai,K. (1981) Isolation and some properties of silver ion-resistant iron-oxidizing bacterium *Thiobacillus ferrooxidans*. *Agricultural and Biological Chemistry* **45**, 2037–2051.

Sugio,T., Domatsu,C., Munakata,O., Tano,T. and Imai,K. (1985) Role of a ferric ion-reducing system in sulfur oxidation of *Thiobacillus ferrooxidans*. *Applied and Environmental Microbiology* **49**, 1401–1406.

Sugio,T., Mizunashi,W., Inagaki,K. and Tano,T. (1987) Purification and some properties of sulfur: ferric ion oxidoreductase from *Thiobacillus ferrooxidans*. *Journal of Bacteriology* **165**, 4916–4922.

Takakuwa,S., Nishiwaki,T., Hosoda,K., Tominaga,N. and Iwasaki,H. (1977) Promoting effect of molybdate on the growth of a sulfur-oxidizing bacterium, *Thiobacillus thiooxidans*. *Journal of General and Applied Microbiology* **23**, 163−173.

Tuovinen,O.H. and Kelly,D.P. (1972) Biology of *Thiobacillus ferrooxidans* in relation to the microbiological leaching of sulphide ores. *Zeitschrift für Allgemeine Mikrobiologie* **12**, 311−346.

Tuovinen,O.H. and Kelly,D.P. (1974a) Studies on the growth of *Thiobacillus ferrooxidans*. IV. Influence of monovalent metal cations on ferrous iron oxidation and uranium toxicity in growing cultures. *Archives of Microbiology* **98**, 167−174.

Tuovinen,O.H. and Kelly,D.P. (1974b) Studies on the growth of *Thiobacillus ferrooxidans*. II. Toxicity of uranium to growing cultures and tolerance conferred by mutation, other metal cations and EDTA. *Archives of Microbiology* **95**, 153−164.

Tuovinen,O.H. and Kelly,D.P. (1974c) Studies on the growth of *Thiobacillus ferrooxidans*. III. Influence of uranium, other metal ions and 2:4-dinitrophenol on ferrous iron oxidation and carbon dioxide fixation by cell suspensions. *Archives of Microbiology* **95**, 165−180.

Tuovinen,O.H., Niemela,S.I. and Gyllenberg,H.G. (1971) Tolerance of *Thiobacillus ferrooxidans* to some metals. *Antonie van Leeuwenhoek* **37**, 489−496.

Tuovinen,O.H., Silver,M., Martin,P.A.W. and Dugan,P.R. (1981) The Agnew Lake uranium mine leach liquors: chemical examinations, bacterial enumeration, and composition of plasmid DNA of iron-oxidizing thiobacilli. In: *Proceedings of the International Conference on Use of Microorganisms in Hydrometallurgy* Hungarian Academy of Sciences, Pecs, pp. 59−69.

Visca,P., Valenti,P. and Orsi,N. (1986) Characterization of plasmids in *Thiobacillus ferrooxidans*. In *Fundamental and Applied Biohydrometallurgy* (eds. R.W.Lawrence, R.M.R.Branion and H.G.Ebner), Elsevier, Amsterdam, pp. 429−441.

Wood,A.P. and Kelly,D.P. (1985) Autotrophic and mixotrophic growth and metabolism of some moderately thermoacidophilic iron-oxidizing bacteria. In *Planetary Ecology* (eds. D.E.Caldwell, J.A.Brierley and C.L.Brierley), Van Nostrand Reinhold, New York, pp. 251−262.

Woods,D.R., Rawlings,D.E., Barros,M.E., Pretorius,I.-M. and Ramesar,R. (1986) Molecular genetic studies on *Thiobacillus ferrooxidans*: The development of genetic systems and the expression of cloned genes. *Biotechnology and Applied Biochemistry* **8**, 231−241.

# CHAPTER 8

# Accumulation and oxidation of metal sulphides by fungi

M. WAINWRIGHT and SUSAN J. GRAYSTON

*Department of Microbiology, University of Sheffield, Sheffield S10 2TN, UK*

## Introduction

The leaching of metals from mineral ores is a process that is generally attributed to the activity of bacteria, while the oxidation of iron, elemental sulphur and mineral sulphides mediated by these bacteria can be used on an industrial scale to extract metals from their ores. Emphasis has been placed on the part played by chemolithotrophic bacteria, most notably species of *Thiobacillus*, in the leaching process, although heterotrophic species can also participate to a lesser extent in the oxidation of metal sulphides (Cole, 1979). The possible involvement of fungi in leaching metals from their sulphide ores has, on the other hand, been neglected. The aim of this chapter is to describe recent work from the authors' laboratory which shows that fungi can oxidize reduced forms of sulphur, including metal sulphides. During the course of these experiments it was noted that fungi possess a remarkable ability to adsorb particulate materials onto their mycelium, including insoluble metal sulphides and elemental sulphur, thereby removing them from solution. Although this phenomenon may be related to the ability of a fungus to oxidize sulphides, a wide range of other particulate materials, for example clays, carbon black and ferric hydrate, are also adsorbed to fungal mycelium so that the process cannot be exclusively related to sulphur oxidation. This adsorptive phenomenon may be used industrially to remove particulates from solution, to clarify waste effluents or to recover valuable particulates such as colloidal gold (Wainwright *et al.*, 1986).

## Bacterial leaching

Before discussing the role of fungi in metal sulphide oxidation it is desirable to consider how bacteria oxidize and solubilize heavy metal sulphides. Metals can be dissolved from insoluble minerals by (i) the direct result of microbial metabolism or (ii) the indirect effects of metabolic products. The main reactions involved in extractive metallurgy are sulphur and mineral sulphide oxidations, the overall reactions for which can be expressed by the equation

$$MS + 2O_2 \xrightarrow{\text{microorganism}} MSO_4$$

where MS is the metal sulphide and M is a divalent metal.

The products of this reaction are usually soluble metal ions and sulphuric acid. As a result, the reaction can be viewed in terms of metal release, which, although industrially useful, may lead to undesirable acidification if it occurs generally in the environment. An example of such detrimental effects occurs when metal sulphides are released from metallurgical plants into the surrounding soil, where they are acted upon by sulphide-oxidizing microorganisms leading to the release of potentially toxic metals and sulphuric acid.

Some bacteria are capable of directly oxidizing mineral sulphides, a process which is associated with attachment to the mineral surface. Such direct attack generally involves specific iron- and sulphur-oxidizing bacteria, the sulphide moiety generally being oxidized to provide energy for chemolithotrophs, with the result that the surface of the mineral becomes visibly pitted.

The solubilization of pyrite involves the following reactions:

$$FeS_2 + 3\tfrac{1}{2} O_2 + H_2O \rightarrow FeSO_4 + H_2SO_4 \tag{1}$$
$$2FeSO_4 + \tfrac{1}{2}O_2 + H_2SO_4 \rightarrow Fe_2(SO_4)_3 + H_2O \tag{2}$$
$$FeS_2 + Fe_2(SO_4)_3 \rightarrow 3FeSO_4 + 2S \tag{3}$$
$$2S + 3O_2 + 2H_2O \rightarrow 2H_2SO_4 \tag{4}$$

Reactions (2) and (4) involve bacterial activity, while the remainder do not. The solubilization of chalcopyrite involves the following reactions:

$$2CuFeS_2 + 8\tfrac{1}{2} O_2 + H_2SO_4 \rightarrow 2CuSO_4 + Fe_2(SO_4)_3 + H_2O \tag{5}$$
$$CuFeS_2 + 2Fe(SO_4)_3 \rightarrow CuSO_4 + 5FeSO_4 + 2S \tag{6}$$

The first reaction involves bacteria, while the sulphur formed can be oxidized to sulphuric acid by reaction (4) shown above.

Indirect oxidation of mineral sulphides involves the participation of the $Fe^{3+}$ ion, either alone or in combination:

$$MS + 2Fe^{3+} + H_2 + 2O_2 \rightarrow M^{2+} + 2Fe^{2+} + SO_4^{2-} + 2H^+ \tag{7}$$
$$Fe_2(SO_4)_3 + FeS_2 \rightarrow 3FeSO_4 + 2S \tag{8}$$

Reaction (8) occurs anaerobically, while reaction (7) is an aerobic process. If iron-oxidizing bacteria are present, then the ferrous sulphate formed in reaction (8) can be oxidized to the ferric state, thereby establishing an oxidation−reduction cycle. Indirect mechanisms of sulphide oxidation can thus be summarized as the chemical reaction of Fe(III) with the sulphide mineral, followed by the regeneration of the Fe(III) by bacteria.

Microorganisms clearly play a major role in the degradation of sulphide minerals. The principal acid-generating microorganism associated with mineral leaching is *Thiobacillus ferrooxidans* which is capable of oxidizing both reduced iron and sulphur compounds in acidic conditions. To a lesser extent, bacteria, fungi, algae and even protozoa may be involved in mineral leaching reactions in nature, although such a diverse assemblage of microorganisms has yet to be put to direct use in industrial leaching processes. Heterotrophic or mixotrophic microorganisms, including species of

*Metallogenium* and *Gallionella*, also appear to be involved in these processes although their importance is not clear.

If fungi are to participate in sulphide leaching then they must be able to participate in reactions mediated by bacteria. In particular they should be capable of directly oxidizing sulphur, metal sulphides and also Fe(II) to Fe(III). Fungi are not usually associated with these processes, but there is evidence that shows that they can participate in such oxidation reactions.

## Sulphur oxidation by fungi

When it became apparent to microbiologists in the early part of this century that the oxidation of reduced sulphur compounds was a microbial process, a search was soon begun for the organisms involved. Lipman *et al.* (1921) and then Waksman and Joffe (1922) showed that sulphur oxidation in soils was mediated by a species of chemolithotrophic colourless sulphur bacteria which they called *Thiobacillus thiooxidans*. This bacterium was shown to be able to fix carbon dioxide autotrophically and gain energy by the oxidation of elemental sulphur to sulphate. Subsequent studies revealed a number of obligate and facultative chemolithotrophic thiobacilli including *T. ferrooxidans*, which gains energy from the oxidation of both iron and sulphur. In the course of the initial enrichment studies for thiobacilli, fungi were often found as contaminants, leading Joffe (1922) to conclude that these organisms play an important role in sulphur oxidation, prior to the involvement of bacteria. Studies by Abbott (1923) and Armstrong (1921) confirmed that fungi could oxidize elemental sulphur and thiosulphate *in vitro*, but little emphasis was placed upon the potential role of fungi in sulphur oxidation until recently when a look at their role was undertaken (Wainwright, 1978).

A wide range of fungi, including species of yeasts, have now been shown to be capable of sulphur oxidation (Wainwright, 1988). Most of the recent work has been devoted to studies on inorganic forms of reduced sulphur although some *Mortierella* species can also oxidize organic sulphide compounds to corresponding sulphoxides (Holland and Carter, 1982). Grayston (1987) showed, however, that *Aspergillus niger, Fusarium solani, Trichoderma harzianum* and *Mucor flavus* were all incapable of oxidizing dimethyl sulphide or dimethylsulphoxide when these were provided in culture medium (0.1% w/v) as the sole carbon source. However, inorganic metal sulphides are oxidized to a limited extent by fungi (Wainwright and Grayston, 1986).

The ability of fungi to oxidize sulphur is not restricted to one taxonomic or physiological group, but occurs widely across all classes amongst, for example, wood-decomposing species, ectotrophic mycorrhizas (Grayston and Wainwright, 1987a) and thermophilous species (Wainwright, 1984a).

Fungi that are obtained from culture collections, and which have apparently had no previous exposure to reduced sulphur, can oxidize the element, so it seems that the ability does not need to be acquired by exposure to sulphur. Fungi that are particularly active sulphur oxidizers include *A. niger* and *Aspergillus flavus, F. solani* and *Trichoderma* species. It should be emphasized, however, that the rates of sulphur oxidation achieved by fungi *in vitro*, while comparable to those achieved by other heterotrophs, fall far short of the rates of oxidation seen when thiobacilli are grown in culture.

The polythionate pathway is usually considered to be the means by which fungi oxidize sulphur because of the appearance of tetrathionate and thiosulphate as well as sulphate in culture media:

$$S^0 \rightarrow S_2O_3^{2-} \rightarrow S_4O_6^2 \rightarrow SO_4^{2-}$$

However, some of these oxyanions may be by-products rather than true intermediates or may even be products of abiotic oxidation reactions. Thiosulphate can be oxidized by fungi to tetrathionate and sulphate, while tetrathionate is oxidized directly to the end product. Sulphur reduction products such as hydrogen sulphide and thiols may also appear in media in which fungi are growing with reduced forms of sulphur (Wainwright and Killham, 1980), products which themselves can be oxidized (Skerman et al., 1957). Although little is known about the biochemistry of fungal sulphur oxidation, it appears to be an enzymatic process both in filamentous fungi (Killham et al., 1981) and in the yeast *Rhodotorula* (Kurek, 1983).

It is unclear what benefits fungi gain from oxidizing sulphur, although they may be capable of growth chemolithoheterotrophically by gaining energy from thiosulphate oxidation while utilizing low concentrations of carbon (Grayston and Wainwright, 1987b). Sulphur oxidation may confer other advantages, including the production of thiols, hydrogen sulphide and polythionates which by complexing heavy metals can reduce their toxicity to fungi (Wainwright and Grayston, 1983). Further details about fungal sulphur oxidation are given in two recent reviews (Wainwright, 1984b, 1988).

**Sulphide oxidation by fungi**

While an increasing number of studies are appearing on the ability of fungi to oxidize sulphur, few studies on sulphide oxidation by fungi have been reported, and as a result little is known about the process. We have recently shown that A.niger can oxidize the sulphides of copper, zinc and lead but not cadmium sulphide to sulphate (*Table 1*). The fungus grew in the form of mycelial spheres onto which the metal sulphides (with the exception of cadmium sulphide) were absorbed. As a result of adsorption the metal sulphide particles were eventually completely removed from the medium, thereby making it clear. This adsorption may have been an active process necessary for sulphide oxidation since it is known that, in an analogous way, sulphur must be attached to the surface of T.thiooxidans before it can be oxidized (Vogler and Umbreit, 1941). A.niger and M.flavus both adsorbed metal sulphides onto their hyphae (*Figure 1*). Cadmium and lead sulphides tended to be adsorbed to the hyphae in clumps (*Figure 2*). Both fungi produced abundant chlamydospores in the presence but not in the absence of the metal sulphides. As chlamydospores are often produced by fungi in response to unfavourable environmental conditions (Ram, 1952; Nickerson and Mankowski, 1953), this observation would indicate that metal sulphides present a somewhat unfavourable environment for the growth of this fungus.

In some cases elemental sulphur was added to the medium together with the sulphides to see if it would increase the rate of sulphide oxidation. Here both elemental sulphur and the sulphides were adsorbed onto the mycelial spheres and the medium was again cleared. In general, larger amounts of sulphate also appeared in the medium when sulphur was present (*Table 1*). In the presence of elemental sulphur cadmium sulphide was

**Table 1.** Effect of growth of *A.niger* on the concentration of sulphur oxyanions, heavy metals, thiols and pH of the medium containing heavy metal sulphides alone and with elemental sulphur

| Metal sulphide | Initial pH | Final pH | $S_2O_3^{2-}$ ($\mu$gSml$^{-1}$) | $S_4O_6^{2-}$ ($\mu$gSml$^{-1}$) | $SO_4^{2-}$ ($\mu$gSml$^{-1}$) | Thiols ($\mu$mol ml$^{-1}$) | Metal ion ($\mu$g ml$^{-1}$) |
|---|---|---|---|---|---|---|---|
| CdS + S$^0$ | 6.7 | 3.7 | ND | ND | 208.4 ($\pm$50.5) | $-5.9$ ($\pm$0.0) | $-48.1$ ($\pm$0.4) |
| CdS alone | 6.7 | 3.2 | ND | ND | $-26.0$ ($\pm$8.4) | $-4.2$ ($\pm$0.8) | $-44.1$ ($\pm$11.3) |
| CuS + S$^0$ | 3.0 | 2.7 | 1.6 ($\pm$2.7) | 294.4 ($\pm$53.0) | 601.0 ($\pm$45.7) | $-1.6$ ($\pm$1.2) | $-25.9$ ($\pm$10.4) |
| CuS alone | 3.0 | 2.4 | ND | ND | 191.2 ($\pm$30.3) | 8.3 ($\pm$2.1) | $-36.2$ ($\pm$7.6) |
| ZnS + S$^0$ | 6.7 | 3.9 | 29.3 ($\pm$6.5) | ND | 354.7 ($\pm$103.3) | 6.9 ($\pm$2.0) | 0.3 ($\pm$3.6) |
| ZnS alone | 6.7 | 4.3 | ND | ND | 89.1 ($\pm$16.1) | 2.0 ($\pm$1.9) | 1.7 ($\pm$1.1) |
| PbS + S$^0$ | 6.7 | 2.3 | 22.4 ($\pm$4.1) | ND | 211.0 ($\pm$67.2) | 5.9 ($\pm$0.1) | $-212.8$ ($\pm$4.8) |
| PbS alone | 6.7 | 2.6 | 28.1 ($\pm$10.2) | ND | 248.5 ($\pm$78.1) | 4.1 ($\pm$0.9) | $-194.9$ ($\pm$32.6) |

Values ($\pm$SD) are expressed as means of triplicates in excess of uninoculated control values; $-$ indicates less than control value (14-day incubation). Sulphate was determined turbidimetrically (Hesse, 1971); thiosulphur and tetrathionate colorimetrically (Nor and Tabatabai, 1976) and thiols as described by Ellman (1959).

ND, not detected.

**Figure 1.** Adsorption of PbS (0.1% w/v) by *M.flavus* from distilled water. (**a**) A control comprising PbS in water; (**b**) dispersion of PbS by *M.flavus* after exposure; (**c**) shows adsorption of PbS by the fungus after 7 days.

adsorbed onto the mycelial spheres of *A.niger*, the medium became clear and excess sulphate was produced (*Table 1*). From SEM studies, it is clear that in the presence of elemental sulphur cadmium sulphide was adsorbed over the surface of the hyphae, and was no longer adsorbed as discrete clumps. It appears, therefore, that the oxidation of elemental sulphur and metal sulphides is linked to the ability of *A.niger* to adsorb these compounds onto its hyphae. Small amounts of thiosulphate were formed by *A.niger* during oxidation of lead sulphide and the sulphides of lead, copper and zinc where elemental sulphur was present (*Table 1*). *A.niger* also formed tetrathionate whilst oxidizing copper sulphide in the presence of elemental sulphur. Sulphide oxidation by *A.niger* was also associated with a marked decrease in the pH of the medium (*Table 1*). The production of extracellular thiols by *A.niger* in media containing heavy metals is notable because of the ability of these compounds to reduce the toxicity of heavy metals to fungi (Ross and Old, 1973).

*Trichoderma harzianum* oxidized copper sulphide to sulphate and also adsorbed the metal sulphide onto its surface. The other metal sulphides were neither adsorbed nor oxidized to sulphate, again suggesting that these processes are in some way linked.

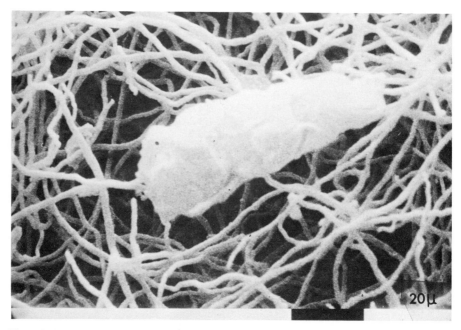

**Figure 2.** Scanning electron microscopy of hyphae of *A.niger* grown in shaking culture containing PbS (0.01% S, w/v; 7 days at 25°C) showing adsorption of clumps of the sulphide.

Thiosulphate and tetrathionate were produced by *T.harzianum* more consistently than by *A.niger*, when oxidizing the metal sulphides (*Figure 3*). Sulphate concentrations in medium in which *T.harzianum* was growing with metal sulphides were, however, lower than those found with *A.niger* (*Figure 3*). Since the amount of sulphate assimilated by these fungi was not determined it is unclear whether *A.niger* was more efficient at oxidizing the sulphides than *T.harzianum*, or merely that it assimilated less of the ion from solution. Fungi have a high requirement for sulphate when growing in nutrient-rich media (Raistrick and Vincent, 1948).

Thiols were produced by *T.harzianum* when oxidizing zinc, lead and copper sulphides (*Figure 4*) but only in small amounts, reflecting the poorer adsorption and apparent oxidation of the metal sulphides compared to *A.niger*. Hydrogen sulphide (detected by blackening of lead acetate paper) was produced by *A.niger* and *T.harzianum* during growth with some of these metal sulphides. This gas was generated by *T.harzianum* only during growth with copper sulphide, where a large excess of sulphate was produced. Hydrogen sulphide was produced by *A.niger* in the presence of all the metal sulphides, with or without elemental sulphur, with the exception of zinc and cadmium sulphides alone, where little or no excess of sulphate was produced (*Table 1*). The formation of this gas appears therefore to be linked with the production of excess sulphate. Sciarini and Nord (1943) suggested that the production of hydrogen sulphide from elemental sulphur resulted from its use as a hydrogen acceptor in alcoholic fermentation of hexoses and pentoses.

The amount of free metal occurring in medium in which *A.niger* and *T.harzianum* grew with heavy metal sulphides is shown in *Table 1* and *Figure 5* respectively. Small

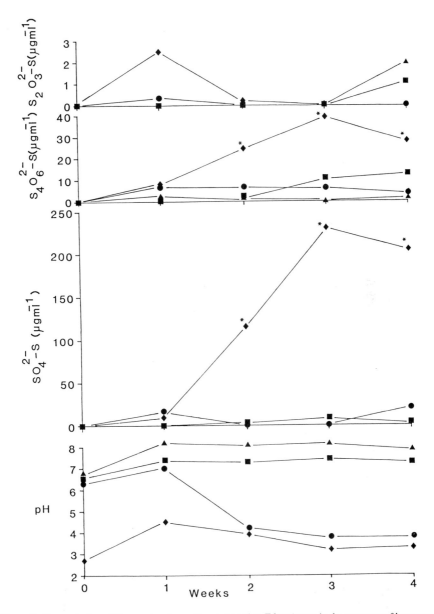

**Figure 3.** Concentration of S-oxyanions in medium supporting *T.harzianum* in the presence of heavy metal sulphides, ●—●, CdS; ♦—♦, CuS; ■—■, PbS; ▲—▲, ZnS. Values are expressed as concentration in excess over uninoculated controls and are means of triplicates. *Significant increase over uninoculated control, $P = 0.05$.

increases in the concentrations of lead (*T.harzianum* only) and zinc (for both fungi) occurred. However, cadmium was the only metal that showed consistent, although not always statistically significant, increases in medium in which both *T.harzianum* grew in the presence of heavy metal sulphides. In most cases, fungal growth led to a reduction

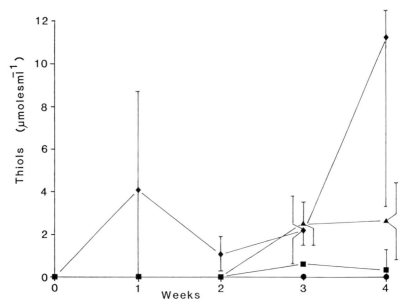

**Figure 4.** Concentration of thiols in medium supporting *T.harzianum* in the presence of heavy metal sulphides (symbols as for *Figure 3*). Means of triplicates ± standard deviation.

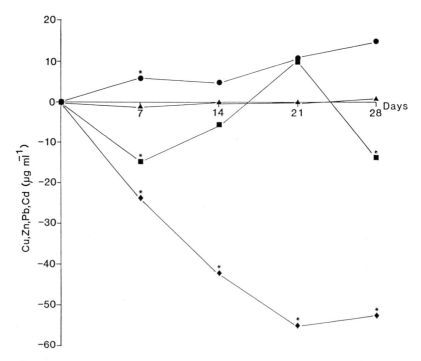

**Figure 5.** Free metal concentration in media supporting *T.harzianum* in the presence of heavy metal sulphides (symbols as for *Figure 3*).

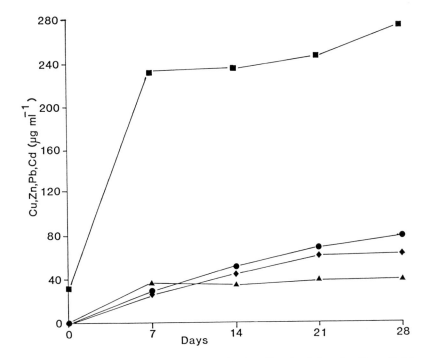

**Figure 6.** Release of metals from sulphides in uninoculated medium (symbols as for *Figure 3*). Means of triplicates, standard deviation never exceeded ± 10% of mean.

in the free metal ion concentration in the medium when compared to uninoculated controls (*Table 1, Figure 5*). Both *T.harzianum* and *A.niger* were clearly capable of removing metals from solution by either an absorption or an adsorption process (Duddridge and Wainwright, 1980). Three processes may account for the observations reported here, namely (i) metal sulphide − wall adsorption, (ii) metal − wall adsorption or (iii) metal uptake. Such adsorption processes may have been less effective in cases where metal ion concentrations increased in the medium (e.g. cadmium, *Figure 5*) or, alternatively, rapid sulphide oxidation might have resulted in a concentration of metal ion which saturated the available metal-adsorption sites. Fungi are known to be able to adsorb soluble metal ions from solution, a process related to the cell-wall constituents, notably chitin, of some filamentous fungi. Tsezos and Volesky (1982) studied the mechanism of biosorption of uranium and thorium in detail and linked it to the chitin in the cell-wall. Keller *et al.* (1983) found that *Penicillium digitatum* also adsorbed uranium from aqueous solution but that this process did not require living hyphae. They tested fungal wall-related biopolymers for the ability to bind uranium and found chitin, cellulose and modified celluloses were active. By this process of adsorption both living and dead fungal mycelium can convert soluble aqueous metal salts into an insoluble form and the metals can then be recovered by filtration or centrifugation and incineration. When grown in high-carbon media, fungi can also produce a polysaccharide (Slodki and Cadmus, 1978) which could also bind the metal or metal sulphide and thereby protect fungi from heavy metal toxicity. *Figure 6* shows that metal sulphides can be solubilized merely by shaking them in media. Such a release of free metal will obviously affect the ability of fungi to grow in the presence of metal sulphides. Adsorption of metals and their sulphides to fungal mycelium might provide a means of enabling fungi to avoid the toxic effects of the heavy metals (Garcia-Toledo *et al.*, 1985). The ability of *T.harzianum* and *A.niger* to produce thiosulphate and tetrathionate should provide an additional source of protection, since the presence of these ions can lead to a reduction in the toxicity of heavy metals to fungi (Wainwright and Grayston, 1983). Similarly, the production of thiols by *A.niger* should help protect fungi against the toxic effects of heavy metals (Ross and Old, 1973).

In a similar experiment to the one described here, the fungi were again grown in batch culture, but the metal sulphides were placed in dialysis tubing in the medium rather than being free in the medium. This method was used by Cole (1979) to study metal sulphide solubilization by heterotrophic soil bacteria. The dialysis tubing prevented the metal sulphides from adhering to the fungal filaments. However, none of the metal sulphides were solubilized or oxidized, again suggesting that the role of adsorption is important in the process of oxidation. Torma *et al.* (1974) and Torma and Sakaguchi (1978) found that the rate of metal sulphide oxidation by *T.ferrooxidans* depended on the solubility product of the metal sulphides involved; the higher the solubility of the metal sulphide, the higher the rate of oxidation. It was suggested therefore that the first step in metal sulphide oxidation is the dissociation of the subtrates. However, from these results it would appear that metal sulphide oxidation by fungi does not follow this pattern. The rate of sulphide oxidation by *A.niger* increased in the order cadmium sulphide < zinc sulphide < copper sulphide < lead sulphide and for *T.harzianum* was in the order zinc sulphide < cadmium sulphide < lead sulphide < copper sulphide. The order of increasing solubility products is copper sulphide < cadmium sulphide < lead sulphide < zinc sulphide. Therefore, there would appear to be no correlation between the solubility product and metal sulphide oxidation in fungi.

It is unlikely that *A.niger* and *T.harzianum* could be used to leach metals from sulphide ores unless the metals were ultimately recovered from the harvested mycelium. *A.niger* might, on the other hand, be used to recover heavy metal sulphide and other particulates from waste waters, where the ability of the fungus to adsorb metal sulphide particles without releasing large amounts of toxic metal ions into solution would obviously be advantageous.

## Conclusions

Based on the results discussed above, there would seem to be little industrial potential for using fungi to leach metals from the sulphide ores, especially since chemolithotrophic bacteria are so efficient in this respect. However, a patent (Richardson, 1981), describes how sulphide ores can be treated by enzymatic action of yeasts to solubilize sulphur and release metals into solution. Yeast spores are apparently mixed in a leaching tank with a carbon source (sucrose), and copper ore, containing 26% (w/v) copper and 30% (w/v) zinc. After 24 h the solution contained 9.5 g Cu $l^{-1}$ and 7 g Zn $l^{-1}$ at pH 2.0.

Fungi may play an important role in the solubilization of metal sulphides in the environment, bringing about an increase in free, and potentially toxic, metal ions, as well as leading to soil acidification.

# References

Abbott,E.J. (1923) The occurrence and action of fungi in soils. *Soil Science* **16**, 207−216.

Armstrong,G.M. (1921) Studies in the physiology of the fungi−sulphur nutrition, the use of thiosulphate as influenced by hydrogen ion concentration. *Annals of the Missouri Botanical Garden* **8**, 237−248.

Cole,M.A. (1979) Solubilization of heavy metal sulphides by heterotrophic soil bacteria. *Soil Science* **127**, 313−317.

Duddridge,J.E. and Wainwright,M. (1980) Heavy metal accumulation by aquatic fungi and reduction in viability of *Gammarus pulex* fed $Cd^{2+}$ contaminated mycelium. *Water Research* **14**, 1605−1611.

Ellman.G.L. (1959) Tissue sulphydryl groups. *Archives of Biochemistry and Biophysics* **82**, 70−77.

Garcia-Toledo,A., Babich,A. and Stotzky,G. (1985) Training of *Rhizopus stolonifer* and *Cunninghamella blakesleeana* to copper: cotolerance to cadmium, cobalt, nickel and lead. *Canadian Journal of Microbiology* **31**, 485−492.

Grayston,S.J. (1987) Sulphur Oxidation and Nitrification by Fungi *in vitro* and in Soils. Ph.D. Thesis, University of Sheffield.

Grayston,S.J. and Wainwright,M. (1987a) Sulphur oxidation by soil fungi including some species of mycorrhizae and wood rotting Basidiomycetes. *FEMS Microbial Ecology* **53**, 1−8.

Grayston,S.J. and Wainwright,M. (1987b) Fungal sulphur oxidation: effect of carbon source and growth stimulation by thiosulphate. *Transactions of the British Mycological Society* **88**, 213−219.

Hesse,P.R. (1971) *A Textbook of Soil Chemical Analysis.* John Murray, London.

Holland,H.L. and Carter,I.M. (1982) The mechanisms of sulphide oxidation by *Mortierella isabellina* NRR1 1957. *Canadian Journal of Chemistry* **60**, 2420−2425.

Joffe,J.S. (1922) Preliminary studies on the isolation of sulfur bacteria from sulfur floats-soil composts. *Soil Science* **13**, 161−172.

Keller,M.G., Macki,D., Feldstein,H., Galun,E., Siegel,S.M. and Siegel,B.Z. (1983) Removal of uranium (VI) from solution by fungal biomass and wall related polymers. *Science* **219**, 285−286.

Killham,K., Lindley,N.S. and Wainwright,M. (1981) Inorganic sulfur oxidation by *Aureobasidium pullulans*. *Applied and Environmental Microbiology* **42**, 629−631.

Kurek,E. (1983) An enzymatic complex active in sulphite and thiosulphate oxidation by *Rhodotorula* sp. *Archives of Microbiology* **143**, 277−282.

Lipman,J.G., Waksman,S.A. and Joffe,J.S. (1921) Oxidation of sulfur by soil micro-organisms. *Soil Science* **12**, 475−489.

Nickerson,W.T. and Mankowski,Z. (1953) A polysaccharide medium of known composition favouring chlamydospore formation in *Candida albicans*. *Journal of Infectious Diseases* **92**, 20−25.

Nor,Y.M. and Tabatabai,M.A. (1976) Extraction and colorimetric determination of thiosulfate and tetrathionate in soils. *Soil Science* **122**, 171−175.

Raistrick,H. and Vincent,J.M. (1948) Studies on the biochemistry of micro-organisms. A survey of fungal metabolism of inorganic sulphates. *Biochemical Journal* **43**, 90−99.

Ram,C.S.V. (1952) Soil bacteria and chlamydospore formation in *Fusarium solani*. *Nature (London)* **170**, 889.

Ross,I.S. and Old,K.M. (1973) Thiol compounds and the resistance of *Pyrenophora avanae* to mercury. *Transactions of the British Mycological Society* **60**, 301−310.

Richardson,F.J. (1981) US Patent 4256485, 11 October, 1979.

Sciarini,L.J. and Nord,F.F. (1943) Elementary sulfur as a hydrogen acceptor in dehydrogenations by living Fusaria. *Archives of Biochemistry and Biophysics* **3**, 261−267.

Skerman,V.B.D., Dermentjev,S. and Skyring,G.W. (1957) Deposition of sulfur from hydrogen sulphide by bacteria and yeast. *Nature (London)* **197**, 742.

Slodki,M.E. and Cadmus,M.C. (1978) Production of microbial polysaccharides. *Advances in Applied Microbiology* **23**, 19−54.

Torma,A.E. and Sakaguchi,H. (1978) Relation between the solubility product and the rate of metal sulphide oxidation by *Thiobacillus ferrooxidans*. *Journal of Fermentation Technology* **56**, 173−178.

Torma,A.E., Legault,G., Kouglou Moutzakis,D. and Ouellet,R. (1974) Kinetics of bio-oxidation of metal sulphide. *Canadian Journal of Chemical Engineering* **52**, 575−577.

Tsezos,M. and Volesky,B. (1982) The mechanism of thorium biosorption by *Rhizopus arrhizus*. *Biotechnology and Bioengineering* **24**, 955−969.

Vogler,K.G. and Umbreit,W.W. (1941) The necessity for direct contact in sulphur oxidation by *Thiobacillus thiooxidans*. *Soil Science* **51**, 331−337.

Wainwright,M. (1978) A modified sulphur medium for the isolation of sulphur oxidizing fungi. *Plant and Soil* **49**, 191−193.

Wainwright,M. (1984a) Sulphur oxidation by some thermophilous fungi. *Transactions of the British Mycological Society* **83**, 721−724.

Wainwright,M. (1984b) Sulphur oxidation in soils. *Advances in Agronomy* **37**, 349−396.

Wainwright,M. (1988) Inorganic sulphur oxidation by fungi. In *Nitrogen, Phosphorus and Sulphur Utilization by Fungi* (eds. L.Boddy, R.Marchant and D.J.Read), Cambridge University Press, Cambridge, pp. 71−88.

Wainwright,M. and Grayston,S.J. (1983) Reduction in heavy metal toxicity towards fungi by addition to media of sodium thiosulphate and sodium tetrathionate. *Transactions of the British Mycological Society* **81**, 541−546.

Wainwright,M. and Grayston,S.J. (1986) Oxidation of heavy metal sulphides by *Aspergillus niger* and *Trichoderma harzianum. Transactions of the British Mycological Society* **86**, 269−272.

Wainwright,M., Grayston,S.J. and De Jong,P. (1986) Adsorption of insoluble compounds by mycelium of the fungus *Mucor flavus. Enzyme and Microbial Technology* **8**, 597−600.

Wainwright,M. and Killham,K. (1980) Sulphur oxidation by *Fusarium solani. Soil Biology and Biochemistry* **12**, 555−558.

Waksman,S.A. and Joffe,J.S. (1922) Micro-organisms concerned in the oxidation of sulfur in the soil. II. *Thiobacillus thiooxidans* a new sulfur oxidizing organism isolated from the soil. *Journal of Bacteriology* **11**, 239−256.

## INDEX